Coronary Care Manual

A practical guide to the management of acute cardiac problems and their subsequent follow-up

J. NOLAN

Senior Registrar
Department of Cardiology
Killingbeck Hospital Leeds

and

D. P. LIPKIN

Consultant Cardiologist
Department of Cardiology
Royal Free Hospital London

Oxford New York Tokyo
OXFORD UNIVERSITY PRESS
1993

Oxford University Press, Walton Street, Oxford OX2 6DP

Oxford New York Toronto
Delhi Bombay Calcutta Madras Karachi
Kuala Lumpur Singapore Hong Kong Tokyo
Nairobi Dar es Salaam Cape Town
Melbourne Auckland Madrid

and associated companies in
Berlin Ibadan

Oxford is a trade mark of Oxford University Press

Published in the United States
by Oxford University Press Inc., New York

A catalogue record for this book is available from the British Library

Library of Congress Cataloging in Publication Data

Nolan J. (James)
Coronary care manual : a practical guide to the management of acute cardiac problems and their subsequent follow-up / J. Nolan, D. P. Lipkin.
(Pocket medical reference series) (Oxford medical publications)
Includes bibliographical references and index.
1. Coronary heart disease—Handbooks, manuals, etc. 2. Coronary Disease—therapy—handbooks. I. Lipkin, D. P. (David P.)
II. Title. III. Series. IV. Series: Oxford medical publications.
[DNLM: 1. Coronary Care Units—handbooks. WG 39 N788c]
RC685.C6N63 1992 616.1'23–dc20 92–49665

ISBN 0–19–262315–X (h/b)
ISBN 0–19–262314–1 (p/b)
Typeset by Footnote Graphics, Warminster, Wiltshire

Printed in Great Britain
by Dotesios Printers Ltd, Trowbridge, Wiltshire

Foreword

Keith A. A. Fox

Duke of Edinburgh Professor of Cardiology, Edinburgh, Scotland

Few areas in medicine have undergone as dramatic a change as that seen in acute cardiac care over the past decade. We have moved from damage limitation and arrhythmia management to complex haemodynamic support, coronary thrombolysis, and mechanical interventions. For the junior physician grappling with the breadth of medical experience, the intensity of modern cardiac care presents a daunting challenge. This *Coronary care manual* provides a clear and logical guide for the junior physician and a useful reference for experienced medical and nursing personnel working with cardiac patients. The clear guidelines and succinct text provide a reassuring reference amid the intensity and complexity of modern acute care. Necessarily in a text of this length, issues have been simplified but further reading lists are provided for more comprehensive discussion of specific areas. This concise handbook of acute cardiac care is a valuable pocket reference, encompassing practical guidelines in the light of the major advances which have taken place over the past 10 years.

Preface

This short book is aimed mainly at medical staff training in coronary care units, though we hope that coronary care nursing staff and non-specialist general physicians will also find it useful. It assumes a basic knowledge of cardiology, and familiarity with routine investigations such as electrocardiograms and chest X-rays.

Most acute cardiac problems are managed by relatively inexperienced members of staff, who often make their decisions about patient management in the early hours of the morning after a long weekend on-call with no ready source of advice to turn to. We hope that this short book will fit easily into a coat pocket, and provide that advice in an accessible format.

The book deals with cardiac problems seen commonly on coronary care units and thus principally with the major cause of death in the United Kingdom, ischaemic heart disease. Guidelines for the subsequent management of these patients following discharge from the coronary unit are also provided.

This book has been kept as short as possible and is not a comprehensive reference work. For more detailed discussion, standard large textbooks of cardiology and relevant publications should be consulted. A list of relevant references is provided at the end of each chapter and in a guide to further reading at the end of the book.

Brief guidelines on performing common practical procedures are included. The best way to learn, however, is to be taught by an experienced colleague, and this chapter is meant to be a supplement to this form of teaching and not a replacement.

Useful information, such as algorithms for the diagnosis and treatment of common conditions and frequently used drug regimes, are grouped together at the end of the book for rapid, easy reference.

The management of acute cardiac problems is a constantly evolving area. Blank pages have been included in the text to enable doctors to add their own notes and updates. We would stress that the treatment regimes outlined in this book are guidelines, and that therapy must be tailored to each individual patient's needs. Recommended drug regimes can change, and the dosage of unfamiliar drugs and their contraindications should be checked in the BNF or manufacturer's literature before they are administered.

Many of the areas covered in this book are still somewhat controversial, but we hope that we have provided sensible guidelines with which most physicians will agree. Inevitably some physicians will hold differing viewpoints. We would welcome any comments and criticisms, and would hope to incorporate improvements in future editions.

Leeds J.N.
London D.P.L.
January 1993

Acknowledgements

Thanks go to Julie Smith, Dr Neil Uren, Dr Stuart Pringle and Professor Keith Fox, who read the manuscript and made innumerable helpful suggestions, Mary Sweeney and Carmel Moran who typed the manuscript, and Sister Rachel Alton who provided the majority of the illustrations. The book is dedicated to our parents who provided help, support, and advice over many years.

Contents

Abbreviations

A+E	accident and emergency unit
APTT	activated partial thromboplastin time
BP	blood pressure
b.p.m.	beats per minute
CCU	coronary care unit
CNS	central nervous system
CO_2	carbon dioxide
COPD	chronic obstructive pulmonary disease
CT	computerized tomography
DC	direct current
DVLC	Driver and Vehicle Licensing Centre
DVT	deep venous thrombosis
ECG	electrocardiogram
EMD	electromechanical dissociation
FBC	full blood count
GA	general anaesthetic
GI	gastrointestinal
GTN	glyceryl trinitrate
IV	intravenous
LDH	lactate dehydrogenase
LFT	liver function tests
PTCA	percutaneous transluminal coronary angioplasty
SVT	supraventricular tachycardia
TFT	thyroid function tests
TIA	transient ischaemic attack
U+E	urea and electrolytes
VF	ventricular fibrillation
VT	ventricular tachycardia
WPW	Wolff–Parkinson–White

CHAPTER 1

Pathophysiology of acute cardiac syndromes

Background

Ischaemic heart disease is the most common cause of mortality in the United Kingdom, accounting for 30 per cent of all deaths. The common acute cardiac ischaemic syndromes seen in hospital practice are myocardial infarction and unstable angina. There are currently around 240 000 acute myocardial infarcts a year in the United Kingdom.

Although there are many possible causes of cardiac ischaemia, it is most commonly due to coronary artery luminal narrowing caused by the development of coronary atheroma. Major risk factors predisposing to the development of coronary atheroma are:

- male sex
- increasing age
- family history
- smoking
- hyperlipidaemia

- hypertension
- diabetes mellitus

Acute cardiac ischaemic syndromes represent a continuous spectrum of disease with the same underlying cause: disruption of a stable atheromatous plaque to form an unstable lesion with superadded thrombus formation. Following disruption and fissuring of a plaque platelet deposition, activation of the clotting cascade and release of vasoactive substances occurs. The end result is narrowing and eventual occlusion of the coronary artery, which may be intermittent. This leads to unstable angina or myocardial infarction, depending on the duration of vessel occlusion, the size of the artery, the site of occlusion, and the presence or absence of collateral vessels.

Unstable angina

In unstable angina, plaque disruption is not usually as extensive as in patients with myocardial infarction, and total occlusion of the ischaemia-related artery is uncommon initially. Episodes of rest pain with associated S–T-segment changes develop and may resolve as the morphology of the coronary artery alters. The process may progress and become irreversible, leading to total coronary occlusion and myocardial infarction in approximately 20 per cent (depending on the population studied) of patients over the next three months. Twelve-month mortality is similar to that of patients with acute myocardial infarction.

Non-Q-wave infarction

This is a syndrome intermediate between unstable angina and Q-wave infarction. Around 25 per cent of all patients admitted with acute myocardial infarction do not develop Q-waves. In the early stages of non-Q-wave infarction, vessel occlusion may be total. Occlusions may occur distally, collateral blood flow is often good, and the time to spontaneous reperfusion is relatively short. Less myocardium is jeopardized and the infarct is usually partial-thickness only. It should be noted that, although non-Q-wave infarcts are usually smaller than Q-wave infarcts with a consequent lower early in-hospital mortality, the incidence of reinfarction is high due to the persistence of an underlying unstable lesion. Overall mortality at one year is similar to that of Q-wave infarction.

Q-wave infarction

In this condition, prolonged occlusion of a major coronary artery occurs. Collateral supply to the myocardium distal to the occlusion is usually poor and a full-thickness area of myocardium infarcts. A large area of myocardium is usually involved, resulting in an extensive infarction. Such extensive full-thickness infarcts are associated with heart failure, ventricular arrhythmias, and a poor prognosis.

Knowledge of the underlying mechanisms of acute ischaemic syndromes has led to the rational use of **aspirin** (for its antiplatelet effect), **nitrates** (to reverse coronary spasm and reduce intraventricular wall tension and therefore myocardial oxygen consumption), **anticoagulants** (to minimize thrombus formation on unstable plaques), **beta-blockers** (to reduce myocardial oxygen demand during an ischaemic episode), and **thrombolytic agents** (to lyse intracoronary thrombi which have formed on unstable plaques).

Key points

- Ischaemic heart disease is the commonest cause of death in the UK.

- Plaque fissuring, with consequent local activation of the clotting system and vasospasm, is the cause of unstable angina and acute myocardial infarction.
- Unstable angina, non-Q-wave infarction, and Q-wave infarction represent a continuous spectrum of disease, with a similar 12-month mortality.

Further reading

Ambrose, J. A., Winters, S. L., Arora, K. K., *et al.* (1986). Angiographic evolution of coronary artery morphology in unstable angina. *J. Am. Coll. Cardiol.*, **7**, 472–8.

Davies, M. J., Woolf, N., and Robertson, W. B. (1976). Pathology of acute myocardial infarction with particular reference to occlusive coronary thrombi. *Br. Heart J.*, **38**, 659–64.

Davies, M. J. and Thomas, A. C. (1985). Plaque fissuring—the cause of acute myocardial infarction, sudden ischaemic death, and crescendo angina. *Br. Heart J.*, **53**, 363–73.

DeWood, M. A., Spores, J., Notske, R., *et al.* (1980). Prevalence of total coronary occlusion during the early hours of transmural myocardial infarction. *N. Eng. J. Med.*, **303**, 897–902.

DeWood, M. A., Stifter, W. F., Simpson, C. S. *et al.* (1986). Coronary arteriographic findings soon after non-Q wave myocardial infarction. *N. Eng. J. Med.*, **315**, 417–23.

Hutter, A. M., DeSanctis, R. W., Flynn, T., and Yeatman, L. A. (1981). Non-transmural infarction: a comparison of hospital and late clinical course of patient with that of matched patients with transmural anterior and transmural inferior myocardial infarction. *Am. J. Cardiol.*, **48**, 595–602.

Madigan, N. P., Rutherford, B. D., and Frye, R. L. (1976). The clinical course, early prognosis, and coronary anatomy of subendocardial infarction. *Am. J. Med.*, **60**, 634–41.

Roberts, W. C. (1972). Coronary arteries in fatal acute myocardial infarction. *Circulation*, **45**, 215–30.

Rose, G. (1991). Epidemiology of atherosclerosis. *Br. Med. J.*, **303**, 1537–9.

Sherman, C. T., Litvach, F., Grundfest, W., *et al.* (1986). Coronary angioscopy in patients with unstable angina pectoris. *N. Eng. J. Med.*, **315**, 913–19.

Silver, M. D., Baroldi, G., and Mariani, F. (1980). The relationship between acute occlusive coronary thrombi and myocardial infarction studied in 100 consecutive patients. *Circulation*, **61**, 219–27.

NOTES

NOTES

NOTES

CHAPTER 2

The coronary care unit

The purpose of the coronary care unit

The aim of a CCU is to concentrate patients with life-threatening cardiac disorders in a well organized area of the hospital staffed by appropriately trained doctors and nurses. This concentration of patients and resources allows ready access to equipment for diagnosing and treating these conditions and their complications.

Approximately 50 per cent of patients with an acute myocardial infarction will die in the community within one or two hours of the onset of their first symptoms. The outlook for these patients can only be improved by community resuscitation schemes and primary-prevention initiatives.

Prior to the thrombolytic era, 15 per cent or more of patients who survived to reach hospital died before discharge. In-hospital mortality can now be reduced by as much as 50 per cent with early therapeutic intervention using **thrombolytic agents**, **aspirin**, **beta-blockers**, and **nitrates**.

The majority of deaths in patients with myocardial infarction who survive to reach hospital occur in the first 24 hours. Patients at particular risk of early death are the elderly, females, those with previous or large myocardial infarcts, and those who develop acute heart failure. Approximately 10 per cent of patients with acute myocardial infarction who are admitted to a CCU will develop a life-threatening ventricular arrhythmia. There is always a risk of sudden life-threatening ventricular arrhythmias in the first few post-infarct hours, even in a seemingly well patient. These should be reversible if treated promptly and correctly. Many life-threatening ventricular arrhythmias occur in patients with well preserved ventricular function and the prognosis is good in these patients if they receive appropriate treatment. Where life-threatening ventricular arrhythmias occur in patients with impaired ventricular function and overt heart failure, the chance of successful resuscitation is much lower.

Following discharge from the CCU most patients do well, with 90 per cent surviving the first year. Those patients most likely to die in their first post-infarct year can now be identified by a series of clinical and investigational features. Medical intervention in these patients may improve their long-term prognosis (see Chapter 14).

General guidelines for admission

All units will have their own individual policies for admission to the CCU; at present it is our policy to admit:

- *all* patients with suspected acute MI
- *all* patients with unstable angina
- selected patients with arrhythmias or complex cardiological problems (i.e. post-cardiac arrest patients from other wards or patients who develop severe acute heart failure) that are likely to benefit from the intensive monitoring and treatment environment of the CCU

Age policies vary, but in general if the patient is fit enough to be resuscitated should cardiac arrest supervene, then admission to the CCU is appropriate.

Key points

- The CCU is an area where staff and equipment are concentrated to optimize the treatment of patients with acute cardiac conditions.
- 50 per cent of patients with an acute myocardial infarction die before reaching hospital.
- In-hospital mortality can now be substantially reduced with modern treatment.
- Most in-hospital deaths occur in the first 24 hours. This is the time when you need to be most vigilant.
- Most patients who survive their time on the CCU do well, with 90 per cent surviving at least 1 year.
- Life-threatening ventricular arrhythmias occurring in the absence of severe impairment of ventricular function have a good prognosis with appropriate treatment. When they complicate severe ventricular impairment, prognosis is less favourable.

Further reading

Armstrong, A., Duncan, B., Oliver, M. F., *et al.* (1972). Natural history of acute coronary heart attacks. A community study. *Br. Heart J.*, **34**, 67–80.

Colling, A., Dellipiani, A. W., Donnaldson, R. J., and McCormack, P. (1976). Teesside coronary survey. An epidemiological study of acute attacks of myocardial infarction. *Br. Med. J.*, **2**, 1169–72.

Julian, D. G. (1987). Jubilee editorial: the history of coronary care units. *Br. Heart J.*, **57**, 497–502.

Tunstall Pedoe, H., Clayton, D., Morris, J. N., Brigden, W., and McDonald, L. (1975). Coronary heart attacks in East London. *Lancet*, **ii**, 833–8.

NOTES

NOTES

NOTES

CHAPTER 3

Diagnostic guidelines

Acute myocardial infarction

The diagnosis of acute myocardial infarction is made in a patient with cardiac ischaemic pain that occurs at rest, usually lasts for more than 30 minutes, and is accompanied by diagnostic electrocardiographic and enzyme changes.

Clinical features

The pain of cardiac ischaemia is typically

- retrosternal
- severe
- crushing
- radiates to the neck, arms, or back

and is associated with nausea, vomiting, and sweating, related to the release of toxins from injured myocardial cells and autonomic activation.

Remember that:

- The pain can be atypical (epigastric, sited in the neck, arms, or back only, or unusual in character) and, particularly with inferior infarcts, can be difficult to distinguish from dyspepsia

- Elderly and diabetic patients often have minimal or no pain.

Physical examination is unremarkable in most patients with acute myocardial infarction, apart from the presence of anxiety, sweating, and a tachycardia. Signs of heart failure or hypotension occur in patients with extensive ventricular damage. Rhythm abnormalities may be obvious in some patients.

Electrocardiographic changes

The typical changes of acute myocardial infarction are as follows.

- S–T elevation in the leads overlying the area of infarction. This can develop within minutes of the onset of infarction, and may persist for several days. Persistence after the first week is often associated with the development of a ventricular aneurysm.
- Pathological Q-waves in the leads overlying the area of infarction. Q-waves are usually associated with full-thickness myocardial infarction, develop within a few hours of the onset of the infarct, and persist in the long term. Associated with the development of Q-waves is loss of R-wave voltage in affected leads.
- T-wave inversion in the leads overlying the area of infarction. This develops as the S–T elevation subsides, and usually resolves over 7–14 days. The first sign of infarction may be the development of tall-peaked T-waves in the area overlying the infarct zone, but the ECG may be completely normal in the early stages of an acute infarct.

The distribution of ECG changes is indicative of the vessel occluded and the area of myocardium involved:

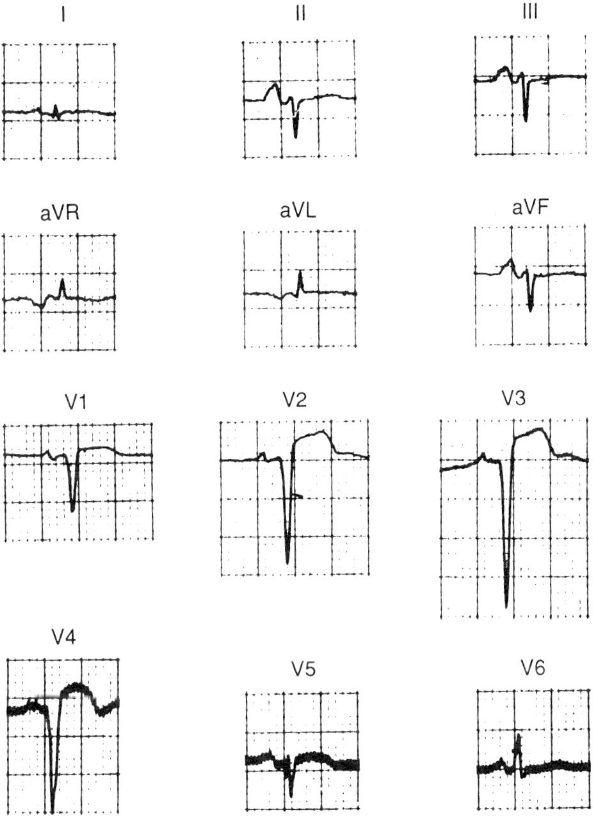

Fig. 3.1 Acute anteroseptal myocardial infarction. There is S–T elevation, T-wave inversion, and Q-waves in V2–V4

- Changes in V2–V4 indicate an anteroseptal infarct, usually due to left anterior descending occlusion (Fig. 3.1).
- Changes in V5–V6 I and aVL indicate a lateral infarct, usually due to circumflex occlusion (Fig. 3.2).

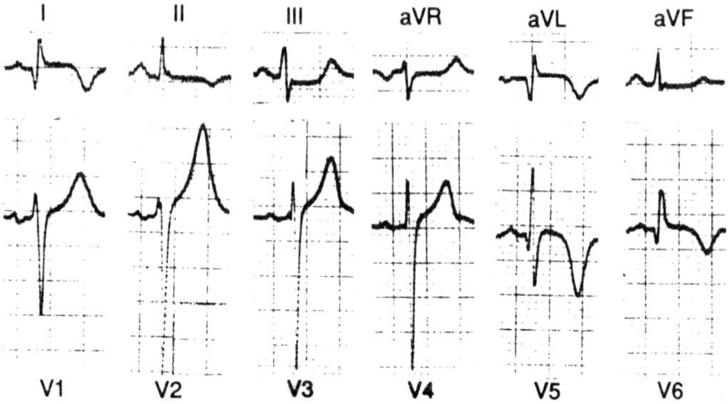

Fig. 3.2 Acute lateral myocardial infarction. There are inverted T waves, resolving S–T elevation and evolving Q waves in V5, V6, I and aVL

- Changes in II, III, and aVF indicate an inferior infarct, usually due to right coronary artery occlusion (Fig. 3.3).
- Tall R-waves in V1 and V2 with associated S–T depression are associated with posterior infarction, usually due to right coronary or circumflex occlusion (Fig. 3.4).

A large infarct is indicated by the presence of widespread changes in several leads, and marked S–T-segment elevation.

Some patients present with a story highly suggestive of intermittent episodes of cardiac ischaemia at rest, but a normal or equivocal ECG. These patients should be

---

Fig. 3.3 Acute inferior myocardial infarction. There is S–T elevation in II, III, and aVF, with a small developing Q-wave in III. There is associated reciprocal S–T depression in the anterior leads, which may represent coexistent left coronary artery disease if it persists after the first few hours of infarction

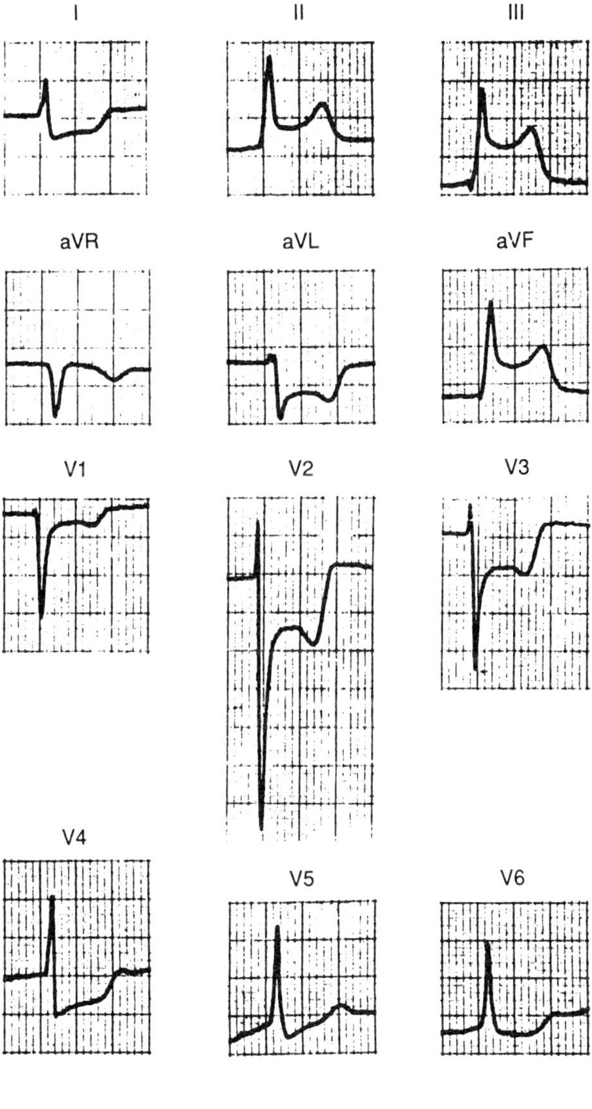
I
II
III
aVR
aVL
aVF
V1
V2
V3
V4
V5
V6

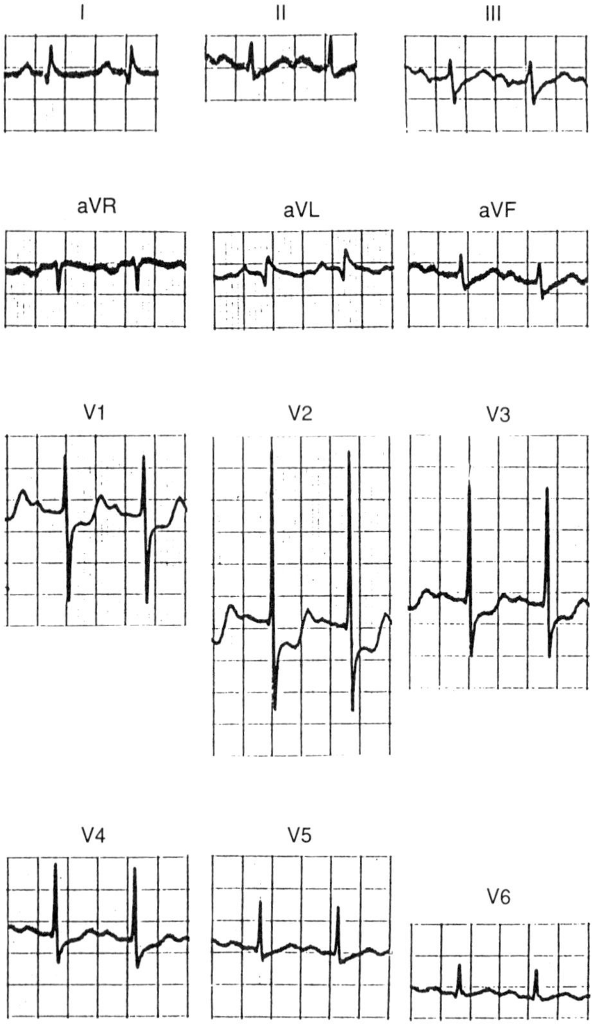
I
II
III
aVR
aVL
aVF
V1
V2
V3
V4
V5
V6

admitted for observation, since they have either unstable angina or an evolving infarct. An ECG should be recorded during any further episodes of pain, or at short intervals in patients with continuing pain, to detect developing diagnostic changes.

A rise in the cardio-specific MB isoenzyme of creatine kinase over the first 48 hours after an episode of chest pain provides the best biochemical evidence of an acute infarct. LDH levels remain elevated for several days after infarction, and may therefore be useful in making a retrospective diagnosis.

Unstable angina

Patients with unstable angina develop episodes of cardiac ischaemic pain (which is similar in site and nature to the pain of stable angina or myocardial infarction) that occur:

- at rest
- on minimal exertion
- is unresponsive to the patient's normal treatment (i.e. unresponsive to GTN in normal doses)
- with increasing frequency
- is prolonged (usually episodes last less than 30 minutes)

Typically, the ECG recorded during an episode of pain shows S–T depression in the territory of the involved coronary artery. Serial ECGs show no S–T elevation or Q-wave development, and cardiac enzymes are not elevated unless the unstable angina progresses to a full-blown infarction.

Fig. 3.4 Tall R-waves in V_1–V_3 with associated S–T depression indicating acute posterior myocardial infarction

Variant angina due to dominant coronary spasm is very rare, and is associated with episodic S–T elevation, T-wave peaking, and pseudo-normalization of T-waves during episodes of pain. Coronary spasm can occur at any time and so symptoms often do not bear any relationship to exertion.

Other causes of chest pain

Non-cardiac chest pain can arise from the following:

- The aorta in acute dissection. The pain of aortic dissection is severe, of sudden onset, tearing in nature, often radiates to the back, and may be associated with hypertension, aortic regurgitation, neurological signs, and pulse deficits.
- The pleura in pneumonia, pulmonary embolism, or pneumothorax. Pain arising from the pleura is unilateral, sharp and stabbing, and worse with inspiration. There may be associated signs of pneumonia, pulmonary embolus, or deep venous thrombosis. In spontaneous pneumothorax, there may be central chest pain with few auscultatory signs. In a patient with COPD, atypical chest pain, and dyspnoea in the absence of ECG evidence of acute myocardial ischaemia, pneumothorax is a strong possibility and a chest X-ray is vital to rule out the presence of air in the pleural space.
- The upper GI tract in oesophageal reflux or peptic ulcer disease. Dyspeptic pain arising from the upper GI tract is usually burning in nature, may have a clear relationship to posture or food, and is often relieved by antacids. Oesophageal pain may, however, be very similar to the pain of cardiac ischaemia. Pain due to ischaemia of the inferior surface of the heart often mimics dyspepsia.

- The pericardium in pericarditis. Pericardial pain is retrosternal, sharp, eased by sitting forward, and may worsen with inspiration. It is commonly seen post-infarction, or in a young adult with acute viral pericarditis. A pericardial rub is common, although it may be intermittent. Diagnostic concave S–T elevation is usually present (Fig. 3.5) in patients with viral pericarditis.
- The bones and muscles in musculoskeletal disorders. Musculoskeletal chest pain is usually unilateral, localized, and sharp. It is exacerbated by movement or local pressure. There may be a clear history of trauma.
- The skin in acute determatological conditions. Acute skin conditions can, rarely, produce chest pain. The unilateral pain of shingles precedes the rash and may confuse the unwary.
- Outside the chest cavity. Referred pain from the gall-bladder or neck. Pain due to gall bladder disease will usually show a clear relationship to food, and may be associated with abdominal signs. Pain referred from the cervical or thoracic spine will have features of musculoskeletal chest pain.

Key points

- Acute myocardial infarction is easily diagnosed when typical pain coexists with clear-cut ECG changes. A normal ECG does not exclude the presence of significant coronary artery disease.
- Pain and ECG changes may be atypical. If you are in doubt about the diagnosis, admit the patient for further evaluation.

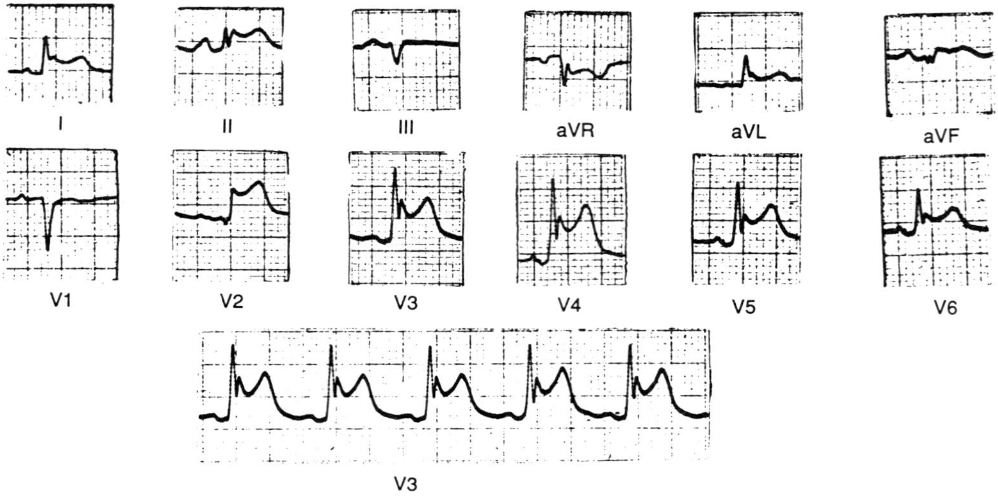

Fig. 3.5 Acute viral pericarditis, showing widespread concave S–T elevation in chest and limb leads

- There are many non-cardiac causes of chest pain. Careful history-taking will allow differentiation in most cases.

Further reading

De Belder, M., French, J., and Camm, A. J. (1989). Electrocardiographic diagnosis of myocardial infarction. *Hosp. Update*, **15**, 671–87.

Lancet (1987). Non-specific chest pain. *Lancet*, **i**, 958–60.

Lee, T. H. and Goldman, L. (1986). Serum enzyme assays in the diagnosis of acute myocardial infarction. *Ann. Intern. Med.*, **105**, 221–33.

Rude, R. E., Poole, W. K., Muller, J. E., *et al.* (1983). Electrocardiographic and clinical criteria for recognition of acute myocardial infarction based on analysis of 3,697 patients. *Am. J. Cardiol.*, **52**, 936–42.

Timmis, A. D. (1990). Early diagnosis of acute myocardial infarction. *Br. Med. J.*, **301**, 301, 941–2.

NOTES

NOTES

NOTES

CHAPTER 4

Initial assessment and management of acute myocardial infarction

Assessment

If a patient has a history suggestive of acute myocardial infarction or unstable angina, and exhibits supporting ECG changes, the diagnosis is usually made easily and rapidly. These patients should not be delayed in the accident and emergency department. This wastes time when the patient is at highest risk of sudden unpredictable arrhythmias, and may delay the institution of thrombolytic treatment.

History-taking should be directed towards establishing the diagnosis, relevant past medical history, drug history, the presence of risk factors for ischaemic heart disease, and the presence of contraindications to thrombolytic therapy (see Chapter 5). Detailed social or family history and comprehensive systematic reviews are not necessary at this stage. Having established a diagnosis of acute myocardial infarction and ruled out contraindications, it should be possible to institute thromolytic therapy within 20 minutes of the patient's arrival. Fast-track admission systems have been shown to reduce the delay from initial presentation to administration of thrombolysis by half. If such a system is in operation in your unit, you should familiarize yourself with its guidelines prior to taking up your post.

Full physical examination is not necessary at this stage. Attention should be directed towards assessment of the vital signs, the cardiovascular system, and the exclusion of alternative diagnoses. Full examination should be performed on the CCU.

Any obvious complications, such as left ventricular failure or arrhythmias, should be treated as a matter of urgency (their management is detailed in later chapters). Remember that in this age of thrombolysis and aggressive early intervention in the course of acute myocardial infarction:

Time lost is muscle lost—try to attend A+E to see a patient within ten minutes of being called.

Optimally, you should stay with the patient (having made your diagnosis and initiated appropriate early therapy) during transfer to the CCU. The cardiac rhythm should be continuously monitored and resuscitation equipment should be immediately available during transfer from A+E to CCU.

Venous access

Having made a diagnosis of acute myocardial infarction or unstable angina, venous access should be established with a cannula in an arm vein. This provides ready access to the circulation for the administration of drugs in an emergency situation. Cannulae should be carefully secured to prevent them falling out in an emergency. Intramuscular injections should be avoided, as drug absorption is unpredictable and thrombolytic therapy enhances haematoma formation.

Antiplatlet therapy

Aspirin has unequivocally been shown to reduce the mortality from acute myocardial infarction and unstable angina. It should be given to all such patients as soon as the diagnosis is made in A+E (see Chapter 11 for dosage and contraindications).

Analgesia and antiemetics

Adequate analgesia is vital to relieve distress and anxiety, both of which cause an increase in sympathetic outflow with resultant detrimental effects on cardiac performance and oxygen consumption. Sublingual **GTN** should be given initially unless systolic blood pressure is below 100 mmHg. If rapid relief of pain within a few minutes does not occur, intravenous narcotic analgesia is necessary. The agent of choice is **diamorphine** 2.5 mg intravenously at 1 mg per minute. The dose should be repeated every 10 minutes until adequate analgesia is achieved. Remember to be cautious not to give too large an initial dose to elderly or small patients, particularly if they have coexistent lung disease, to avoid respiratory depression.

Metoclopramide 10 mg intravenously should be given with narcotic analgesics to minimize nausea and vomiting.

Nitrates

An overview of all the major trials of nitrates suggests that they are beneficial in acute myocardial infarction. Nitrates reduce myocardial oxygen consumption by decreasing pre- and afterload, and reverse coronary vasoconstriction, thus helping to limit infarct size and to control episodes of ischaemic pain. Optimally the dose administered should be increased to produce a small fall in blood pressure, since doses that fail to do this are less likely to have a beneficial effect on infarct size.

Nitrates should be administered routinely to all patients with unstable angina (see Chapter 13) and acute myocardial infarction, unless the systolic blood pressure is below 100 mmHg on admission. The first dose should be given in A+E as 600 μg of sublingual **GTN**. On arrival in the CCU, an intravenous nitrate infusion should be commenced (the intravenous rather than other routes of administration is preferable as it allows fine titration of administered dose against haemodynamic effect) using one of the regimes detailed in Chapter 18. Infusions should be started at the minimum recommended dose, and increased every 15 minutes until side-effects occur, systolic BP falls by 20 mmHg, systolic BP drops to 100 mmHg, or excessive tachycardia occurs. Buccal **GTN** (5 mg repeated every 4 hours) is an alternative and effective regime which is cheaper than intravenous nitrates, though it is not as easy to titrate this regime to produce the desired haemodynamic response.

The optimal duration of nitrate therapy is unknown. It is reasonable to continue therapy for 24 hours, then to tail off the dosage, unless evidence of ongoing ischaemia (i.e. continuing chest pain or ECG changes) is present. Nitrate tolerance develops rapidly, and if administration is required after the first 24 hours, consideration of the use of interrupted regimes or a progressive increase in dosage will be necessary.

Investigations to perform on the CCU

On arrival on the CCU, all patients should have the following investigations:

- U+E
- glucose
- FBC
- lipids
- cardiac enzymes
- chest X-ray

U+E, glucose, and FBC results should be obtained as soon as possible and necessary action taken if abnormalities such as severe anaemia or electrolyte deficiencies are present. Lipid levels and cardiac enzyme studies should be sent for routine laboratory analysis, as they will not usually influence management in the first 24 hours.

A chest X-ray should be carried out on the CCU. It is not appropriate to waste time in A+E doing a chest X-ray; it is potentially dangerous to send a patient unattended to X-ray during this vulnerable time. The aim of radiological examination is to assess cardiac size (though this can be unreliable on a portable film), look for early signs of left ventricular failure, and exclude obvious mediastinal widening (a sign of possible aortic dissection). If the diagnosis is in doubt and a dissection is a possibility, further investigation is necessary.

Thrombolytic therapy should not be administered to a patient with a possible aortic dissection. Remember that a chest X-ray that does not show mediastinal widening does not rule out the diagnosis of acute dissection. If a non-cardiac cause of chest pain is a possibility, then the chest X-ray may provide diagnostic information, demonstrating a clinically inapparent pneumothorax or skeletal abnormality.

Oxygen therapy

Administer oxygen in high concentration to all ill patients (i.e. those with left ventricular failure, hypotension, arrhythmias or evidence of continuing myocardial ischaemia) as detailed in Chapter 18. If there is a possibility of CO_2 retention in a patient with a history of COPD who requires oxygen therapy, it is mandatory to check blood gases and administer controlled oxygen therapy (see Chapter 18). Patients who are well on admission and do not develop complications probably do not merit oxygen therapy.

Thrombolysis and beta-blockade

All patients with acute myocardial infarction should be considered for thrombolytic therapy and intravenous beta-blockers, as detailed in Chapters 5 and 6, unless a contraindication exists. Only a few of the contraindications are absolute, and where doubt exists seek senior advice. Some units advocate the initiation of thrombolytic therapy in the A+E department, whereas other units prefer to commence therapy on the CCU. Local factors, such as the distance between A+E and the CCU and average transfer times, will determine the particular policy in operation and you should ascertain this policy at the commencement of your CCU attachment.

Patients who are already being treated with beta-blockers at the time of admission should have them continued unless a persistent bradycardia (less than 50 b.p.m.), hypotension, or acute heart failure develop.

Serum potassium

If serum potassium is below 4.0 mmol/l in the presence of an important arrhythmia (i.e. an arrhythmia that causes hypotension, heart failure, cardiac ischaemia, or distressing symptoms) then potassium supplements should be given:

1. If no central line is in place, give 20 mmol KCl in 100 ml 5% dextrose over 1 hour; this will raise serum potassium by 0.5 mmol/l, and can be repeated as necessary. This often causes pain at the infusion site, but pain can be minimized by slowing the infusion rate.
2. If a central line is in place, 40 mmol of KCl in 100 ml 5% dextrose can be given over 1 hour, raising serum potassium by 1.0 mmol/l.

The serum potassium should be re-checked after an interval of 2–3 hours to ensure that it is maintained above 4.0 mmol/l.

If serum potassium is below 4.0 mmol/l in the absence of an important arrhythmia, then give oral potassium supplements (for example **Slow K**, 3 tablets t.d.s., providing approximately 120 mmol of potassium daily) and re-check potassium after 12–18 hours.

Magnesium supplementation

If an intractable or life-threatening arrhythmia develops in a patient who has been receiving long-term diuretic therapy, magnesium deficiency is a possible aggravating factor. Patients should receive magnesium replacement as well as potassium replacement using the following regime:

1. Give 25 mmol **magnesium chloride** in 20 ml 5% **dextrose** over 3 hours.
2. Give 50 mmol **magnesium chloride** in 200 ml 5% **dextrose** over the next 21 hours.
3. Give 25 mmol **magnesium chloride** in 200 ml 5% **dextrose** over the following 24 hours.

There is preliminary evidence that administration of magnesium to patients with acute myocardial infarction may improve survival. Routine prophylactic use of magnesium cannot be recommended until the results of the ISSIS-4 trial are published.

Hyperglycaemia

If on admission glucose is above 11 mmol/l, then treatment with IV insulin via a sliding scale should be started using the regime given in Chapter 18; the aim is to reduce blood glucose to 7.0–11.0 mmol/l, not to achieve strict euglycaemia. Elevation of blood sugar to 10–15 mmol/l is common. Elevation above this level in a non-diabetic subject is less common; if it occurs you should be alert to the possibility of severe metabolic derangement developing, which will require intensive corrective therapy. You should consult a senior colleague in this situation if you do not feel confidently able to diagnose and treat diabetic emergencies adequately.

Further management of hyperglycaemia after the first 24 hours will depend on the underlying metabolic state. If the blood glucose was not elevated above 15 mmol/l and has returned to normal 24 hours after admission, then no further investigation is necessary. Elevation of blood glucose above 15 mmol/l or for more than 24 hours may be due to underlying undiagnosed diabetes. A glucose tolerance test should be carried out prior to discharge and appropriate dietary or drug therapy instituted if a diagnosis of diabetes is confirmed.

Anxiolytic therapy

All patients should receive appropriate reassurance. For particularly anxious patients, therapy with **diazepam** 5 mg t.d.s. may be helpful in the short term. **Temazepam** 10 mg *nocte* is useful as a hypnotic. Benzodiazepines should not routinely be continued after discharge from hospital.

Key points

- History-taking and physical examination of patients with suspected acute myocardial infarction should be directed towards confirming the diagnosis, ruling out alternative causes of chest pain, and establishing the patient's suitability for thrombolytic therapy.
- In a clear-cut case of acute myocardial infarction without contraindications *do not delay administering thrombolysis*, as maximum benefit is to be gained from early administration.
- Adequate analgesia, antiemetics, aspirin, and sublingual nitrates should all be given in A+E. Thrombolysis can be administered in A+E or CCU according to local guidelines.
- Routine investigations should be arranged on the CCU not in the A+E.

- All patients should be considered for thrombolysis, intravenous nitrates, and intravenous beta-blockade.
- Correction of hypokalaemia and, where appropriate hypomagnesaemia, is vitally important in the control of arrhythmias.

Further reading

Gwilt, D. J., Nattrass, M., and Pentecost, B. L. (1982). Use of low dose insulin infusion in diabetics after myocardial infarction. *Br. Med. J.*, **285,** 1402–4.

Hollenberg, N. K. and Brown, R. S. (1984). Electrolytes and cardiovascular disease. Proceedings of a symposium. *Am. J. Med.*, **77**(5A), 1–10.

Husband, D. J., Alberti, K. G. M., and Julian, D. G. (1983). 'Stress' hyperglycaemia during acute myocardial infarction: an indicator of pre-existing diabetes? *Lancet*, **ii,** 179–81.

ISIS-2 (Second International Study of Infarct Survival) Collaborative Group (1988). Randomised trial of intravenous streptokinase, oral aspirin, both or neither among 17 187 cases of suspected acute myocardial infarction. *Lancet*, **ii,** 349–60.

Jugdutt, B. I. and Warnica, J. W. (1988). Intravenous nitroglycerin therapy to limit myocardial infarct size, expansion and complications: effect of timing, dosage, and infarct location. *Circulation*, **78,** 906–19.

Lancet (1991). Magnesium for acute myocardial infarction. *Lancet*, **338,** 667–8.

Lipkin, D. P. and Reid, C. J. (1988). Myocardial infarction: the first 24 hours. *Br. Med. J.*, **296,** 947–8.

Nordrehaug, J. E. (1985). Malignant arrhythmias in relation to serum potassium in acute myocardial infarction. *Am. J. Cardiol.*, **56,** 20d–23d.

Pell, A. C. H., Miller, H. C., Robertson, C. E., and Fox, K. A. A. (1992). Effect of 'fast track' admission for acute myocardial infarction on delay to thrombolysis. *Br. Med. J.*, **304,** 83–7.

Teo, K. K., Yusef, S., Collins, R., Held, P. H., and Peto, R. (1991). Effects of intravenous magnesium in suspected acute myocardial infarction: overview of randomised trials. *Br. Med. J.*, **303**, 1499–503.

Woods, K. L., Fletcher, S., Roffe, C., Haider, Y. (1992). Intravenous magnesium sulphate in suspected acute myocardial infarction: results of the second Leicester Intravenous Magnesium Intervention Trial (LIMIT-2). *Lancet*, **339**, 1553–8.

Yusef, S. (1988). The use of beta adrenergic agents, intravenous nitrates, and calcium channel blocking agents following acute myocardial infarction. *Chest*, **93**, 25s–85s.

Yusef, S., MacMahon, S., Collins, R., and Peto, R. (1988). Effect of intravenous nitrates on mortality in acute myocardial infarction: an overview of the randomized trials. *Lancet*, **i**, 1088–92.

NOTES

NOTES

NOTES

CHAPTER 5

Thrombolysis

Background

Recent large well-conducted clinical trials have shown unequivocal reduction in mortality in acute myocardial infarction with all of the available thrombolytic agents. There is no evidence that any of the currently available agents are superior. Maximum reduction in mortality occurs when the agents are used early in the course of the infarction. Remember:

Time lost is muscle lost.

The ISIS-2 study of **streptokinase** has shown beneficial effects from treatment given to patients with anterior and inferior infarctions, in all age groups and in either sex. It is probable that patients who present from 12–24 hours after the onset of symptoms with no residual pain and established Q-waves have little to gain from thrombolytic therapy. Most authorities now recommend routine use of thrombolysis, providing no contraindication exists, up to 12 hours after the onset of symptoms. From 12–24 hours thrombolysis is recommended if there is evidence of continuing ischaemia—i.e. persistent pain associated with ECG evidence of evolving infarction. If you are unsure about giving thrombolysis late in the course of an infarct, consult a senior colleague for advice.

Streptokinase is considerably cheaper than the other thrombolytic agents. **Tissue plasminogen activator** (**altepase**) may cause earlier reperfusion than streptokinase but this is not clearly translated into beneficial long-term effects on mortality or residual left ventricular function. Taking this into account, streptokinase is the first choice thrombolytic agent, using tissue plasminogen activator where streptokinase is contraindicated.

The role of **APSAC** (**anistreplase**) and **urokinase** remains to be fully defined. **APSAC** has the advantage that it can be given as a single IV dose over 4 minutes. This ease of administration is offset by its cost and lack of superior efficacy. **Urokinase** is expensive and offers no clear-cut advantages. Combination thrombolytic regimes are being investigated at present in an attempt to combine the advantageous properties of the different agents (i.e. rapid, early action of **tissue plasminogen activator** combined with prolonged effect of **streptokinase**) but comprehensive data are not yet available and these regimes are not recommended for routine clinical use.

The risk of major complications is low. In particular, the excess risk of stroke is negligible even in the elderly population. The ISIS-3 study suggests that the risk of stroke is higher with **tissue plasminogen activator** and **APSAC** than with **streptokinase**. Minor side-effects are relatively common, but seldom serious. Following the use of **streptokinase**, neutralizing antibodies reach significant levels after about five days. These antibodies persist for at least 12 months. Use of **streptokinase** in this period is contraindicated as the antibodies present may neutralize the drug preventing its therapeutic thrombolytic effect, and **tissue plasminogen activator** should be used in these circumstances. Similarly, a recent streptococcal infection will induce antibody formation, and **tissue plasminogen activator** should be used preferentially. Hypotension occurs more commonly following administration of **streptokinase** than **tissue plasminogen activator**. If admission blood pressure is low, **streptokinase** should not be used, as it may cause further hypotension.

Aspirin used in conjunction with **streptokinase** has an additive effect on reducing mortality. Overall vascular mortality post-infarction is reduced by approximately 20 per cent by either therapy alone, and by approximately 40 per cent when they are used together; the maximum reduction in mortality occurs when combination therapy is started early.

Definite ECG changes of infarction should be present before administration of thrombolysis to a patient with ischaemic pain. If doubt exists and the ECG is equivocal, then the patient should receive intensive therapy for unstable angina and the ECG should be repeated every 20 minutes to detect evolving changes. If a clear-cut infarct pattern develops on serial electrocardiography, thrombolytic therapy is indicated. If symptoms settle and the ECG does not evolve, the patient should continue to be treated as having unstable angina. Mortality in patients with chest pain and a normal ECG is very low. Limiting thrombolytic therapy to patients with definite ECG changes of infarction will reduce the risk of inappropriate administration.

As discussed in detail in Chapter 11, **heparin** is not routinely recommended after **streptokinase**, but should be used after **tissue plasminogen activator**. Oral anticoagulants are only recommended for patients at high risk of thromboembolic complications.

Inclusion criteria for streptokinase

- Presentation within 12 hours of the onset of ischaemic chest pain of at least 30 minutes' duration. From 12–24 hours, treatment is indicated if there is clear-cut evidence of continuing ischaemia or an evolving MI (i.e. persistent chest pain and S–T-segment elevation). Local policy may vary, and you should check with a senior colleague what time limits are in operation on your unit.
- Persistent S–T elevation of at least 2 mm in two chest leads, 1 mm in two limb leads, or changes of true posterior infarction.

Exclusion criteria for streptokinase

- known coagulation disorder
- potential site of GI bleeding (i.e. active peptic ulcer or known oesophageal varices)
- severe hypertension: systolic BP > 200 mmHg and/or diastolic BP > 120 mmHg
- systolic pressure below 90 mmHg on admission (**tissue plasminogen activator** is preferred in hypotensive patients)
- prolonged recent cardiopulmonary resuscitation
- recent bleeding from any site
- history of stroke with residual neurological deficit, or TIA in the last three months
- surgery, major trauma, or head injury in the last month
- concurrent anticoagulant therapy at the time of admission
- proliferative retinopathy: check carefully the retina of a diabetic patient before giving opiate, as the resulting pupillary constriction will render fundoscopy difficult

- pregnancy or menstruation
- treatment with **streptokinase** more than five days and less than 12 months previously
- recent streptococcal infection

Few of these contraindications are absolute. If you are in doubt about a patient's suitability for thrombolysis then consult a senior colleague before commencing thrombolytic therapy.

Remember to exclude the diagnosis of aortic dissection (suspect if pain is very severe, radiates to the back, and is of sudden onset in the presence of risk factors for dissection such as Marfan's syndrome hypertension or suggestive physical signs such as an aortic regurgitant murmur, neurological defects, absent pulses, or a wide mediastinum on chest X-ray), pericarditis (widespread, concave S–T-segment elevation on ECG), and acute abdominal emergency (presence of abdominal signs in a patient with an atypical history).

Administration regime for streptokinase

- Steroids and antihistamines should not be used prophylactically as this has not been shown to decrease the incidence of allergic reactions, and routine use may encourage infarct expansion
- Infuse 1.5 million units of **streptokinase** in 100 ml **normal saline** over one hour.
- Give **aspirin** 160 mg on admission, and 160 mg daily thereafter.

Signs of reperfusion are unreliable, but rapid resolution of S–T elevation is the best routinely available indicator. Arrhythmias are common post-infarction even when thrombolysis is not given, and they cannot be used as a reliable indicator of successful thrombolysis. Reperfusion rates at 90 minutes are around 60–70 per cent for intravenous streptokinase.

Insertion of central venous lines following thrombolysis can be dangerous. If central access is needed for pacing or haemodynamic monitoring, then the internal jugular or femoral venous route should be used preferentially.

Complications and their management

Cerebrovascular events

The overall incidence of stroke is increased by only 0.1 per cent since thrombolytic therapy causes a slight excess of cerebral haemorrhages which is offset by a reduction in cerebral infarcts.

Allergy

Anaphylaxis is rare (0.1 per cent). When it occurs, initial treatment is 1 ml of 1:1000 **adrenaline** as a slow intravenous bolus or intramuscular injection if IV access is not obtainable. 200 mg of **hydrocortisone** and 10 mg of **chlorpheniramine** should also be given, but their onset of action is delayed and they will not reverse the acute cardiovascular collapse that occurs with anaphylaxis.

Mild allergic reactions (such as urticaria) are also uncommon, and can be treated with antihistamines and steroids if necessary.

Haemorrhage

Minor bleeding at venepuncture sites is relatively common (3–4 per cent) but rarely requires any specific therapy other than compression at the bleeding site.

Major haemorrhagic episodes requiring transfusion are rare (0.3 per cent). If a major bleed occurs;

1. Stop **streptokinase** infusion.

2. Give two units of fresh frozen plasma immediately.
3. Give **tranexamic acid** 10 mg/kg by slow IV injection.

Hypotension

Hypotensive reactions to streptokinase infusion can occur. They should be treated initially by tilting the patient's head down, and giving **atropine** 0.6 mg IV if a bradycardia is also present. If the hypotension persists and is clearly associated with the streptokinase infusion, then the drug should be stopped and fluids and inotropes administered cautiously if necessary. **Tissue plasminogen activator** can be substituted for streptokinase in these patients as it is less likely to cause hypotension.

Management after thrombolysis

Management of patients after thrombolysis should be along conventional lines, as outlined for post-infarct care in Chapter 14. Reocclusion rates of 10–20 per cent occur following thrombolysis, but clinically evident reinfarction is less common.

Early PTCA in asymptomatic patients is avoided, as studies to date have not shown it to be of any prognostic benefit. Delayed PTCA at 7–10 days may be useful in the control of post-infarction angina unresponsive to drug therapy. Indications for revascularization following acute myocardial infarction are discussed further in Chapter 14.

If reinfarction occurs in hospital, further thrombolytic therapy should be given, using **streptokinase** if less than 5 days have elapsed since the initial administration. **Tissue plasminogen activator** is preferred after 5 days, to avoid the neutralizing effect of antibodies to streptokinase.

Indications and administration regime for tissue plasminogen activator

Tissue plasminogen activator is used preferentially as first choice thrombolytic agent when the admission systolic BP is less than 90 mmHg or neutralizing antibodies to **streptokinase** are likely to be present. Contraindications to the use of **tissue plasminogen activator** are otherwise identical to those for streptokinase. Complications and efficacy are similar. Tissue plasminogen activator is given as:

1. A 10 mg bolus over 1–2 minutes.
2. An infusion of 50 mg over 1 hour.
3. An infusion of 40 mg over the next 2 hours.

The dose of tissue plasminogen activator may need to be reduced in small patients, and details of dose modification are provided in the manufacturer's guidelines. There is a significant risk of early reocclusion following thrombolysis with tissue plasminogen activator, and **heparin** should be given as detailed in Chapter 11.

Key points

- All the currently available drugs have a similar beneficial effect on mortality.
- **Streptokinase** is the first choice agent since it is the cheapest. **Tissue plasminogen activator** is first-choice agent in certain limited circumstances.
- The risk of major complications is low. Thrombolysis should be given to all patients with acute myocardial infarction in any site, irrespective of age, who present within 12 hours of the onset of symptoms. From 12–24 hours, patients may still benefit if they have evidence of continuing ischaemia or evolving infarction.

- **Aspirin** should be given to all patients. **Heparin** should only routinely be used after tissue plasminogen activator. **Warfarin** should not routinely be used unless a specific indication is present.

Further reading

AIMS Trial Study Group (1988). Effect of intravenous APSAC on mortality after acute myocardial infarction: preliminary report of a placebo controlled clinical trial. *Lancet*, **i,** 545–9.

ASSET study group (1988). Trial of tissue plasminogen activator for mortality reduction in acute myocardial infarction: Anglo–Scandinavian study of early thrombolysis. *Lancet*, **ii,** 525–30.

Erbel, R., Pop, T., Diefenbach, C., and Meyer, J. (1989). Long term results of thrombolytic therapy with and without percutaneous translumenal coronary angioplasty. *J. Am. Coll. Cardiol.*, **14,** 276–85.

GISSI 2 (1990). A factorial randomised trial of alteplase versus streptokinase and heparin versus no heparin among 12,490 patients with acute myocardial infarction.. *Lancet*, **336,** 65–71.

Gruppo Italiano per Lo Studio della Streptochinasi nell' Infarto Miocardico (GISSI) (1986). Effectiveness of intravenous thrombolytic treatment in acute myocardial infarction. *Lancet*, **i**, 397–402.

ISIS-2 (Second International Study of Infarct Survival) (1988). Collaborative group: randomized trial of intravenous streptokinase, oral aspirin, both, or neither among 17,187 cases of suspected acute myocardial infarction. *Lancet*, **ii**, 349–60.

ISIS-3 (Third International Study of Infarct Survival) (1992). Collaborative group: a randomized comparison of streptokinase vs tissue plasminogen activator vs anistreplase, and of aspirin plus heparin vs aspirin alone among 41,299 cases of suspected acute myocardial infarction. *Lancet*, **339,** 753–70.

Simoons, M. L., Betriu, A., Col, J., *et al.* (1988). Thrombolysis with tissue plasminogen activator in acute myocardial infarction: no additional benefit from immediate coronary angioplasty. *Lancet*, **i,** 197–203.

SWIFT (1991). SWIFT trial of delayed elective intervention versus conservative treatment after thrombolysis with anistreplase in acute myocardial infarction. *Br. Med. J.*, **302,** 555–60.

The International Study Group (1990). In-hospital mortality and clinical course of 20,891 patients with suspected acute myocardial infarction randomised between alteplase and streptokinase with or without heparin. *Lancet*, **336,** 71–5.

TIMI Research Group (1988). Immediate versus delayed catheterisation and angioplasty following thrombolytic therapy for acute myocardial infarction. TIMI II A Results. *J.A.M.A.*, **260,** 2849–58.

TIMI study group (1989). Comparison of invasive and conservative strategy after treatment with intravenous tissue plasminogen activator in acute myocardial infarction. Results of the thrombolysis in myocardial infarction (TIMI) phase II trial. *N. Eng. J. Med.*, **320,** 618–26.

Topol, E. J., Califf, R. M., George, B. S., *et al.* (1987). A randomised trial of immediate versus delayed elective angioplasty after intravenous tissue plasminogen activator in acute myocardial infarction. *N. Eng. J. Med.*, **317,** 581–8.

NOTES

NOTES

NOTES

NOTES

CHAPTER 6

Intravenous beta-blockers

Background

Intravenous beta-blockers given early in the course of acute myocardial infarction have been shown to reduce in-hospital mortality by 15 per cent, probably by decreasing the incidence of cardiac rupture. Since thrombolytic therapy may increase the risk of early cardiac rupture, it is logical to use concurrent intravenous beta-blockers to offset this risk. Although data are limited, combination therapy seems to have an additive effect on reducing the rate of early death, with in-hospital mortality rates as low as 5 per cent being achieved with combination therapy in the TIMI II trial. Provided the list of contraindications is closely adhered to, this is a safe and effective therapy and should be considered in all acute infarcts. Patients who are hypertensive on admission (systolic BP of 160 mm Hg or greater) are at greatest risk of cardiac rupture, and are therefore must likely to gain benefit from the use of intravenous beta blockers. **Metoprolol**, **propranolol**, or **atenolol** have all been shown to be efficacious. We recommend using **metoprolol** since it has a relatively short half-life and is cardioselective.

Inclusion criteria

Patients should be considered for IV beta-blockers if:

- presentation is within twelve hours of the onset of symptoms
- definite ECG changes of infarction are present with S–T elevation of at least 2 mm in two chest leads or 1 mm in two limb leads

Exclusion criteria

Patients should be excluded if:

- the resting heart rate is below 60 b.p.m.
- admission systolic BP is below 100 mmHg
- first-, second- or third-degree heart block present
- the patient is receiving concurrent treatment with beta-blockers or **verapamil** at time of admission
- overt left ventricular failure is present
- bronchospasm is present on admission or the patient gives a history of asthma or COPD

Remember to be cautious about giving IV beta-blockers to a patient with a marked resting tachycardia, as this may well be due to heart failure.

Administration regime for metoprolol

Metoprolol is given as an initial IV bolus to control heart rate, followed by oral metoprolol as follows:

1. Give 5 mg **metoprolol** IV slowly over 2 minutes. Stop if heart rate drops to 40 b.p.m. or below, systolic BP drops to 100 mmHg or below, the P–R interval is prolonged to more than 0.26 s, or dyspnoea is aggravated.

2. If heart rate remains above 60 b.p.m. 2 minutes after initial IV bolus of **metoprolol**, give a further slow IV injection of 5 mg, followed by a further 5 mg 2 minutes later if heart rate remains above 60 b.p.m., observing the above cautions, to a maximum total dose of 15 mg.
3. Fifteen minutes after final IV dose, give **metoprolol** 50 mg orally, repeated every 6 hours for the next 48 hours. Withhold oral therapy if heart rate 50 b.p.m. or below.
4. Continue long-term oral beta-blockade unless a contraindication develops. (Chapter 14).

Complications and their management

Serious complications are rare. The most important are:

- Precipitation of left ventricular failure. This should be treated with intravenous loop diuretics initially.
- Bradycardia. If this is prolonged or symptomatic then treat with IV **atropine** 0.6 mg. If this is unsuccessful, an infusion of a beta-agonist such as **dobutamine** can be used. If the above drugs fail, a temporary pacing wire will be necessary.
- Hypotension. This should be treated with **dobutamine** initially. If it is severe, prolonged, and refractory to **dobutamine**, then **glucagon** can be used to reverse beta-blockade (50 μg/kg IV as a bolus, followed by an infusion of 1–5 mg per hour initially).
- Bronchospasm. This should be treated with inhaled beta-agonists such as **salbutamol**.

Other less serious side-effects (cold peripheries, CNS effects, and so on) occur, but seldom require any specific intervention or merit cessation of therapy.

Key points

- Intravenous beta-blockade given in the first 12 hours after the onset of symptoms of myocardial infarction will reduce in-hospital mortality by up to 15 per cent.
- We recommend use of a cardioselective agent with a relatively short half-life such as **metoprolol**.
- The aim of the initial IV dose is to reduce resting heart rate to the range of 40–60 b.p.m.
- Long-term oral beta-blockade should be considered for all patients who do not have a specific contraindication.

Further reading

ISIS-1 (First International Study of Infarct Survivals) Collaborative Group (1986). Randomised trial of intravenous atenolol among 16,027 cases of suspected acute myocardial infarction. *Lancet*, **ii**, 57–66.

Julian, D. G., Chamberlain, D. A., Sandoe, E., *et al.* (1988). Mechanisms for the early mortality reduction produced by beta blockade in acute myocardial infarction. *Br. Heart J.*, **59**, 615–16.

MIAMI trial research group (1985). Metoprolol in acute myocardial infarction (MIAMI). A randomised placebo controlled trial. *Eur. Heart J.*, **6**, 199–226.

Sleight, P. (1986). Use of beta adrenoceptor blockade during and after acute myocardial infarction. *Ann. Rev. Med.*, **37**, 415–25.

Yusef, S., Peto, R., Lewis, J., Collins, R., and Sleight, P. (1985). Beta blockade during and after myocardial infarction: an overview of the randomised trials. *Prog. Cardiovasc. Dis.*, **27**, 335–71.

NOTES

NOTES

NOTES

NOTES

CHAPTER 7

Arrhythmias

General principles

The vast majority of patients on the CCU have ischaemic heart disease and their rhythm disturbance is likely to be related to this. This chapter deals with the treatment of arrhythmias in patients with ischaemic heart disease unless otherwise stated.

- Arrhythmias are common after acute myocardial infarction, occurring in 95 per cent of patients. The commonest rhythm disturbance is increased ventricular ectopic activity.
- Polypharmacy is to be avoided. If first and second line drug therapy fails, consider DC cardioversion as the next step. If the patient develops haemodynamic compromise as a result of the arrhythmia or drug therapy, proceed directly to DC cardioversion.
- Maintaining serum potassium above 4.0 mmol/l is important. Long-term diuretic therapy can induce both potassium and magnesium depletion, and this has been shown to reduce the threshold for developing rhythm disturbances (see Chapter 4 for details of

potassium and magnesium administration in patients with arrhythmias).

- **Avoid negatively inotropic drugs in patients with an initial systolic BP below 100 mmHg, as they may precipitate severe hypotension.**
- DC cardioversion is the treatment of choice for sustained ventricular or supraventricular arrhythmias where the patient is severely haemodynamically compromised or does not respond to alternative forms of treatment. Patients with severe haemodynamic impairment usually have the following features:
  1. systolic BP $<$ 80 mmHg
  2. overt left ventricular failure
  3. poor cerebral perfusion
  4. poor coronary perfusion (i.e. ischaemic chest pain present at rest)
  5. poor renal perfusion (i.e. oliguria present)
  6. poor peripheral perfusion (i.e. cold, clammy peripheries)

Supraventricular tachyarrhythmias

Sinus tachycardia

Sinus tachycardia is common after acute myocardial infarction, as a response to anxiety, pain, or most commonly heart failure. The rate is usually from 100–160 b.p.m. Onset and termination are gradual and the P-wave axis is normal (upright in lead II, inverted in AVR). The QRS complex is narrow. Treatment is directed at the cause of the tachycardia:

- Treat heart failure.
- Ensure adequate analgesia if continuing chest pain is present.

- Minimize anxiety with reassurance and benzodiazepines.

Persistent sinus tachycardia is usually associated with extensive infarction and severe heart failure and implies a poor prognosis.

Supraventricular ectopics

These are recognized by their narrow QRS complex, presence of abnormal P-waves, or their premature position in the cardiac cycle (Fig. 7.1). They are common following infarction and usually require no treatment. Increasingly frequent supraventricular ectopics may herald the onset of atrial fibrillation.

Atrial fibrillation and flutter

Atrial fibrillation and flutter are common post-myocardial infarction, affecting 15 per cent of patients. Their occurrence is associated with increasing age, extensive infarction, and an increased mortality. Fifty per cent of all episodes last less than 30 minutes. Atrial fibrillation or flutter is recognized by the presence of a narrow complex tachycardia, absence of P-waves, the presence of fibrillation or flutter waves, and complete beat-to-beat irregularity in the case of atrial fibrillation (Fig. 7.2). A regular narrow complex tachycardia with a

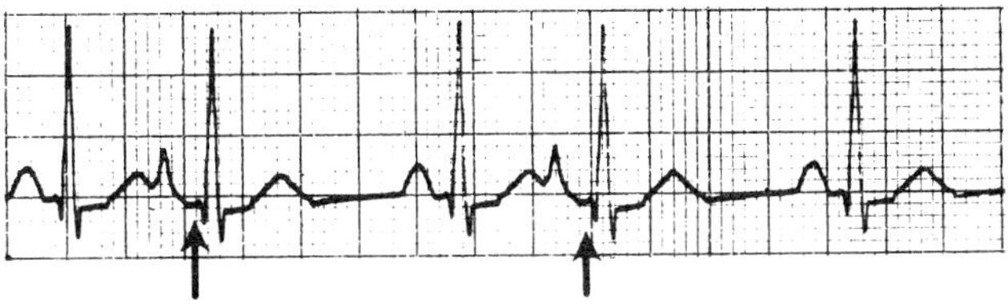

Fig. 7.1 Supraventricular ectopic beats (arrowed). These are premature, narrow complex, and have associated abnormal P-waves

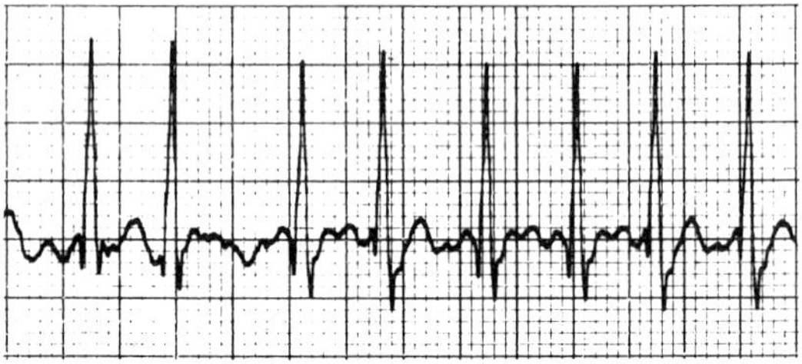

Fig. 7.2 Atrial fibrillation, showing rapid, irregularly irregular ventricular response (180 b.p.m.)

rate of 150 b.p.m. should be considered to be atrial flutter with 2:1 AV block until proven otherwise (Fig. 7.3). If atrial fibrillation or flutter occurs in the absence of a rapid ventricular rate, in an otherwise well patient who is not haemodynamically compromised (i.e. ventricular rate below 130 b.p.m. and systolic BP > 100 mmHg) then no treatment is necessary initially.

If the rhythm disturbance persists for more than 30 minutes, or the ventricular rates rise above 130 b.p.m. then treatment is indicated to reduce excessive myocardial oxygen consumption. First-line treatment in this situation is oral **digoxin**:

1. Give 0.5 mg **digoxin** initially.
2. Give 0.5 mg **digoxin** after 8 hours.
3. Give 0.5 mg **digoxin** after a further 8 hours.
4. Give 0.25 mg **digoxin** daily thereafter.

This regime may need to be modified and lower doses given in elderly patients or those with impaired renal function.

If systolic BP is 90–100 mmHg but the patient is otherwise well (i.e. no angina, cerebration normal, urine output adequate, and good peripheral perfusion) treat with **digoxin** as detailed above, but have a low

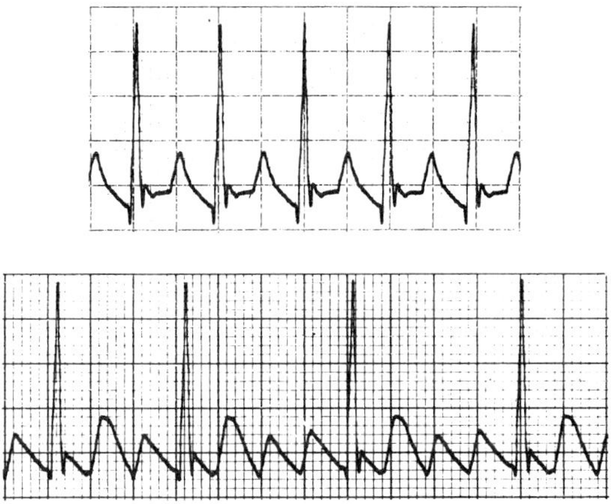

Fig. 7.3 (a) shows a rapid, regular narrow complex tachycardia with a regular rate of 150 b.p.m. in a patient with atrial flutter and 2:1 block, (b) shows the effect of carotid sinus massage which slows the ventricular rate to reveal underlying flutter waves with a rate of 300 b.p.m.

threshold for proceeding to DC cardioversion if the blood pressure falls or myocardial ischaemia develops.

Digoxin should be given orally where possible, as absorption is rapid and complete. If the patient is vomiting or absorption is likely to be impaired then the recommended oral doses can be given intravenously in 100 ml normal saline over 30 minutes for each dose. Attention to correction of hypokalaemia is important when using **digoxin** as a low serum potassium will potentiate toxicity.

If systolic BP is less than 90 mmHg with the onset of atrial flutter or fibrillation, or falls despite digoxin, DC cadioversion is the treatment of choice. (*Note*: atrial flutter is very responsive to DC cardioversion, and will often revert with low energy shocks of only 20–30 J.)

Class I antiarrhythmic drugs should never be administered to patients with atrial flutter or fibrillation unless a nodal blocking agent (i.e. **digoxin**, **verapamil**, or **metaproltol**) is given concurrently, or a paradoxical increase in the ventricular rate may occur. The aim of treatment of atrial flutter or fibrillation should be to achieve a resting ventricular rate of 70 b.p.m.

Paroxysmal SVT

This is not common post-infarction. It is usually seen on the CCU in the context of a young person with a structurally normal heart and a long history of previous attacks. The following discussion relates to these patients, not to the treatment of arrhythmias in patients with ischaemic heart disease.

Paroxysmal SVT is usually due to re-entry circuits within the AV node. Onset is abrupt, rate is 140–220 b.p.m., and the QRS complex is narrow (Fig. 7.4). The arrhythmia is usually well tolerated by the patient.

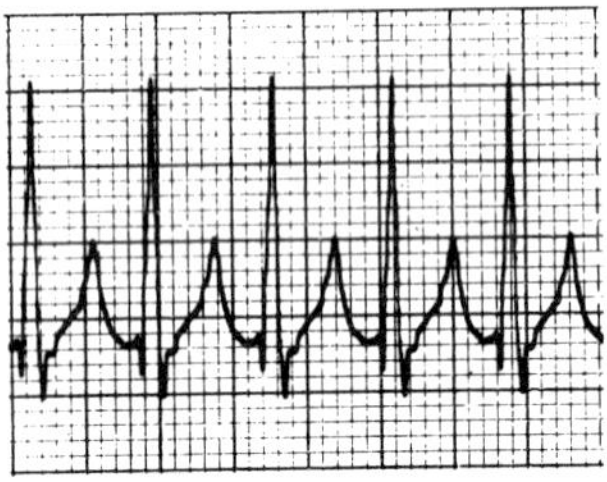

Fig. 7.4 Paroxysmal SVT. Regular narrow complex tachycardia with a rate of 180 b.p.m.

Initial treatment should be with carotid sinus massage or other vagal manoeuvres. If these fail, and the arrhythmia persists, drug therapy is indicated. Intravenous **Adenosine** is now the drug of first choice, given as an initial rapid bolus of 3 mg, followed by further rapid boluses of 6 and 12 mg if the arrhythmia does not terminate. **Adenosine** may cause bronchospasm, and

Verapamil (given in 5 mg boluses over 30 seconds, repeated after 5–10 minutes if necessary up to a maximum of 20 mg) is preferable in patients with asthma or COPD. If the BP falls below 90 mmHg, or the arrhythmia persists despite drug therapy, consider DC cardioversion.

Verapamil should be avoided when:

- initial systolic BP is below 100 mmHg
- the patient has impaired ventricular function or known ischaemic heart disease
- there is any doubt as to the ECG diagnosis, and VT is a possibility
- the patient is receiving concurrent therapy with beta-blockers, as this combination may result in profound bradycardia and hypotension.

In such patients, adenosine is the agent of first choice. Prophylaxis against recurrent attacks of SVT can be with:

- **digoxin** 0.25 mg daily
- **verapamil** in doses of 40–120 mg t.d.s.
- beta-blockers i.e. **atenolol** 50–100 mg daily or **sotalol** 40–160 mg t.d.s. (**sotalol** has both beta-blocking and Class III antiarrhythmic activity)
- **flecainide** in doses of 50–100 mg b.d.
- **disopyramide** in doses of 150–400 mg b.d.

Paroxysmal atrial tachycardia with AV block

In this rhythm disturbance, atrial rate is 140–250 b.p.m., but the ventricular rate is slower due to blocked conduction in the atrioventricular node. The degree of block may be variable and the ventricular rate may therefore be irregular. The QRS complex is narrow.

This arrhythmia usually occurs in patients with structural heart disease and is therefore often poorly tolerated. Over 50 per cent of cases are due to **digoxin** toxicity.

If the patient is receiving **digoxin** this should be stopped and hypokalaemia corrected if present. An urgent **digoxin** level should be requested. Beta-blockers can be used if necessary to slow the ventricular rate. DC cardioversion is contraindicated in a patient with **digoxin** toxicity as it may induce intractable ventricular arrhythmias.

If this rhythm disturbance arises in the absence of **digoxin** therapy, or in a patient whose **digoxin** level is subtherapeutic, then this drug can safely be used to control the ventricular rate if necessary.

Wolff–Parkinson–White syndrome

In patients with Wolff–Parkinson–White syndrome, an accessory pathway connects the atria and ventricles. This allows atrial impulses to bypass the AV node, initiating ventricular activation prematurely. The syndrome is characterized by a short P–R interval, a broad QRS complex due to the presence of a delta wave (Fig. 7.5), and a tendency to paroxysmal tachycardia. The syndrome is congenital and occurs in 1.5/1000 of the population.

The commonest arrhythmia associated with Wolff–Parkinson–White syndrome is paroxysmal SVT. This is normally initiated by an ectopic beat setting up a circus movement, down the AV node and back up the accessory pathway. The ECG during the tachycardia will show a rapid, regular narrow complex SVT with no delta waves, and the diagnosis can only confidently be made from an ECG recorded before or after the tachycardia.

Narrow complex regular SVTs in patients with WPW syndrome are terminated by suppressing conduction through the AV node, thereby interrupting the circus movement of electrical activity from atria to ventricles and back that generates the arrhythmia. **Digoxin** and **verapamil** can accelerate conduction in the bypass tract

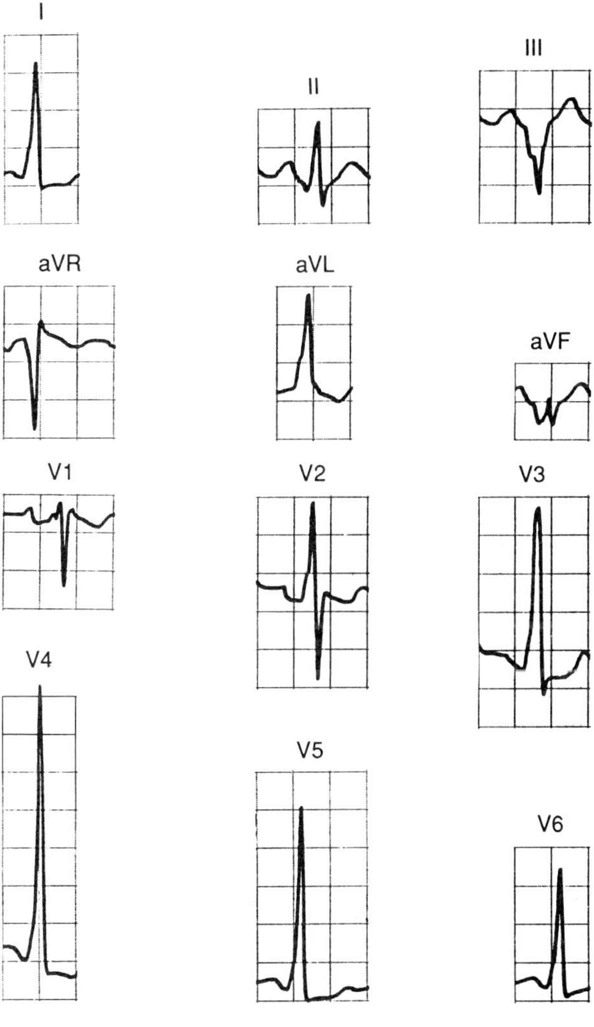

Fig. 7.5 Wolff–Parkinson–White syndrome, showing short P–R interval and delta wave

if atrial fibrilation occurs and should be avoided. The following IV agents are suitable for use in SVT due to WPW:

- **Adenosine** 3–12 mg IV as described on p. 70
- **Metoprolol** 5 mg IV at a rate of 1 mg 1 min repeated after 5 minutes if necessary to a total close of 15 mg.

If the arrhythmia is associated with a systolic BP below 100 mg, DC cardioversion is indicated. If drug therapy with a single agent fails, consult a senior colleague for advice. Consider oral therapy with a beta blocker **disopyramide flecainide** or **amiodarone** for recurrent attacks, but you should be guided by senior colleagues in this difficult area.

Atrial fibrilation occurs in up to 30% of patients with WPW. Conduction from atria to ventricles occurs down the bypass tract causing a rapid irregular broad complex tachycardia. Treatment should be with agents that slow bypass tract conduction (**Disopyramide** 2 mg/kg IV over 5 minutes, **Flecainide** 2 mg/kg IV over 10 minutes or **Amiodarone** 5 mg/kg IV over 15 minutes). If the ventricular rate is very rapid and the minimal interval between successive beats is less than 250 ms, there is a significant risk of sudden death due to VF during a tachycardia, and the patient should be referred for specialist electrophysiological investigation and treatment.

Overdrive pacing for supraventricular tachyarrhythmias

If a temporary pacing wire is *in situ*, then overdrive atrial pacing is an alternative treatment for supraventricular tachycardia. Atrial fibrillation will not respond to overdrive pacing. Major advantages of this form of non-pharmacological therapy are that a general anaesthetic is avoided, and the treatment can be repeated if the arrhythmia recurs.

For overdrive pacing:

1. Place a temporary pacing wire in contact with the endocardium in the right atrium.
2. Pace the right atrium at a rate 20 per cent faster than the tachycardia for 30 seconds. (A special overdrive pacing box capable of rapid pacing rates is necessary.)
3. Terminate the pacing abruptly and normal sinus rhythm will often return. If this fails, repeat the overdrive pacing, but gradually decrease the pacing rate over several minutes to try and capture the ventricles and reduce the rate. Atrial fibrillation may supervene; this usually reverts to sinus rhythm over the next few hours and, even if not, is easer to control than the original SVT or flutter.

Ventricular tachyarrhythmias

Ventricular ectopics

These manifest themselves as broad complex premature beats (Fig. 7.6). They are very common (95 per cent of patients) after acute myocardial infarction and require no specific therapy other than correction of hypokalaemia if present. The concept of complex ventricular ectopic activity being a warning sign, heralding the onset of VF, is unreliable as they are neither sensitive nor specific. Many patients develop VF with no preceding excess ectopic activity, and many patients have high-grade complex ectopic activity but do not develop VF. Routine treatment of ventricular ectopic beats post-infarction is not justified as it has not been shown in clinical trials that such treatment is of any value. Routine use of prophylactic **lignocaine** for all patients with acute

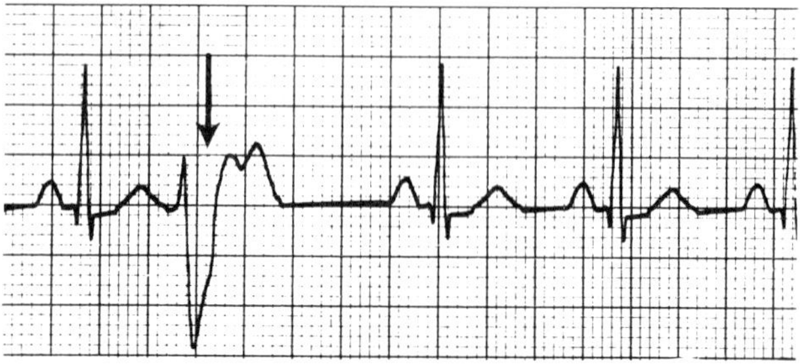

Fig. 7.6 Multifocal ventricular ectopic beats (arrowed) which are premature, broad complex, and of variable morphology

infarction is not advocated. Although this reduces the incidence of ventricular fibrillation, it increases the risk of asystole and therefore has no overall beneficial effect. If ventricular ectopics are very frequent, ensure that the patient is not in pain, treat anxiety with **benzodiazepines**, and maintain plasma potassium above 4.0 mmol/l.

Ventricular tachycardia

This a common life-threatening arrhythmia after acute myocardial infarction, presenting as a rapid (rate above 180 b.p.m.) broad complex (QRS duration greater than 110 ms) tachycardia. There is a high risk of haemodynamic collapse or progression of the arrhythmia to VF. Broad complex tachycardia can be supraventricular in origin if rate-related aberrant conduction, pre-excitation, or pre-existing bundle branch block is present. The differentiation of VT from these other causes of broad complex tachycardia is important since VT requires different treatment, is a medical emergency, and carries a worse prognosis. When faced with a patient with a broad complex tachycardia, remember:

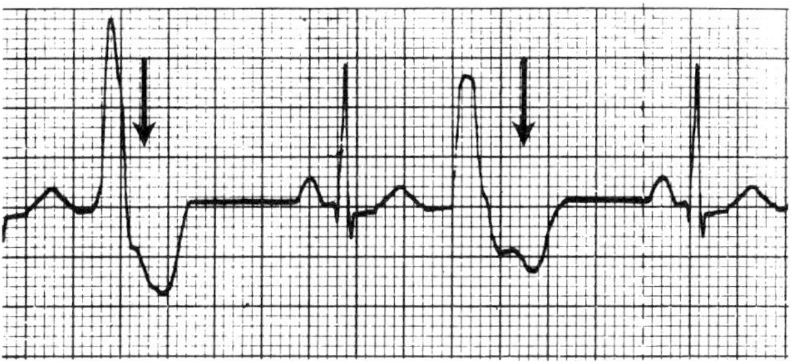

- Most patients on the CCU have ischaemic heart disease, and a broad complex tachycardia is more likely to be VT than SVT.
- If the patient is aged over 35, and has a history of ischaemic heart disease or heart failure, there is a 90 per cent probability that a broad complex tachycardia is VT.
- If you are in any doubt as to the diagnosis, assume it to be VT and treat appropriately.
- Not all patients with VT are acutely unwell. The patient's general clinical state is therefore not a reliable guide to the diagnosis

The presence of two or more of the following ECG criteria suggests that a broad complex tachycardia is VT and not SVT:

- rate greater than 180 b.p.m.
- atrioventricular dissociation (present in 45 per cent of patients with VT)
- fusion or capture beats
- RSR pattern in V1, with the primary R-wave being taller than the secondary R-wave
- deep S-wave in V6

- concordant pattern across the precordial leads—i.e. QRS vector in the same direction V1 to V6
- left-axis deviation (in a patient with a narrow QRS during sinus rhythm)
- QRS complex duration greater than 140 ms (in a patient with a narrow QRS during sinus rhythm)
- regular RR intervals—irregularity of VT is uncommon and due to capture or fusion beats disrupting the regularity of the tachycardia

These criteria allow accurate diagnosis of VT in 95 per cent of cases from the surface electrocardiogram (Fig. 7.7). An algorithm for the ECG diagnosis of broad complex tachycardia is provided in Chapter 18.

Broad complex tachycardias that do not fulfil these criteria are likely to be due to supraventricular tachycardia with aberrant conduction, or tachycardia in the presence of pre-excitation. Where doubt exists as to the diagnosis of a broad complex tachycardia and the patient is not haemodynamically compromised (i.e. systolic blood pressure is at least 100 mmHg), an intravenous bolus of **adenosine** will terminate the majority of supraventricular tachycardias but will have no effect on a ventricular tachycardia. **Adenosine** has a very short half life and is well tolerated in VT. If the tachycardia persists despite **adenosine**, it should be regarded as a VT and treated appropriately. **Adenosine** is given as an initial IV bolus of 3 mg, repeating the bolus every 2 minutes if no effect is noted and intolerable side-effects do not occur, increasing the dose to 6 then 12 mg on each successive occasion. **Adenosine** may cause bronchoconstriction in susceptible individuals, and a history of asthma or obstructive airways disease is a relative contraindication to its use.

Treatment of VT

If the systolic BP is less than 100 mmHg with the onset of the tachycardia, the patient is distressed or poorly perfused, the treatment of choice is DC cardioversion. If the patient is conscious with the arrhythmia, sedate with IV **diazempam** 5–10 mg or administer the shock under a brief GA. If consciousness is lost with the arrhythmia, then a DC shock must be administered immediately.

If the systolic BP is greater than 100 mmHg and the patient is not distressed or poorly perfused give **lignocaine** 1.5 mg/kg IV over 1 minute. If this terminates the arrhythmia commence a **lignocaine** infusion at 4 mg/minute for 30 minutes, 2 mg/minute for two hours, and 1 mg/minute thereafter. If the VT recurs give a further bolus of 1.5 mg/kg of **lignocaine** and increase the infusion rate to 2 mg/minute.

If the initial bolus of lignocaine fails to terminate the VT and the patient remains haemodynamically and clinically stable, give **disopyramide** 2 mg/kg IV to a maximum of 150 mg over 5 minutes. If this terminates the arrhythmia commence a **disopyramide** infusion at 0.4 mg/hour (Chapter 18). If the VT recurs a further IV bolus of disopyramide can be given. The IV dose should not exceed 800 mg in 24 hours or 300 mg in the first hour.

If therapy with **lignocaine** and **disopyramide** fails to terminate the VT proceed to DC cardioversion. If VT recurs during lignocaine or disopyramide infusion despite the measures described above stop the infusion and replace it with **amiodarone** 5 mg/kg IV over 20 minutes followed by an infusion of 1200 mg over 24 hours (Chapter 18).

If a temporary pacing wire is *in situ*, you can attempt to terminate the VT by overdrive pacing. You should pace the ventricle at a rate of 20–30/per minute above the rate of the tachycardia for 10 seconds, then either

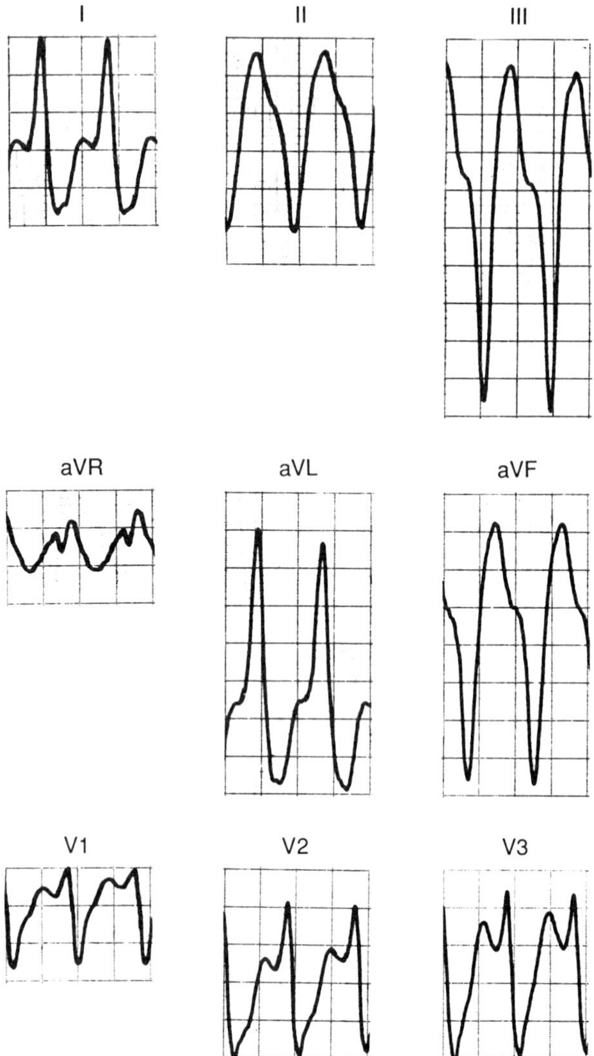
I
II
III
aVR
aVL
aVF
V1
V2
V3

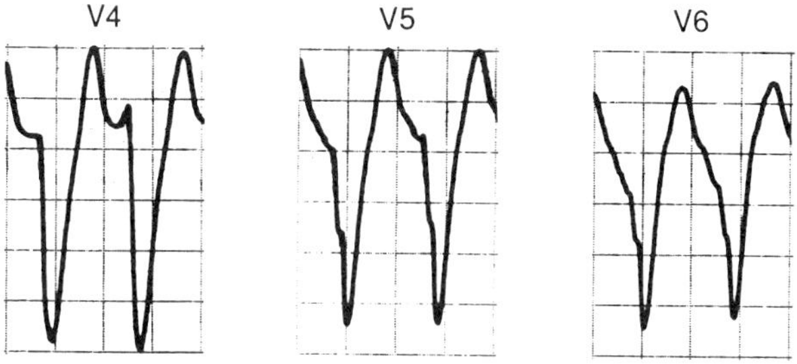

Fig. 7.7 Ventricular tachycardia; there is a broad complex regular tachycardia with ventricular concordance, and left-axis deviation.

terminate the pacing abruptly or gradually decrease the pacing rate. This treatment is only applicable in a haemodynamically stable patient. If the patient is compromised with the onset of the tachycardia or deteriorates during attempts to terminate the arrhythmia with overdrive pacing, proceed directly to DC cardioversion.

After termination of VT, correct any metabolic abnormalities that may be present (check U+E, glucose, and blood gases), check an ECG, and treat any aggravating factors such as heart failure or continuing myocardial ischaemia.

An algorithm for the treatment of VT is provided in Chapter 18.

Ventricular fibrillation

See the section on resuscitation (Chapter 8, pp. 97–104) for details of the treatment of VF.

Idioventricular rhythm

This is a common rhythm disturbance following acute myocardial infarction. It is now seen more commonly since the advent of thrombolytic therapy. It manifests as a broad complex tachycardia with a rate below 120 b.p.m.

(Fig. 7.8) which rarely causes any haemodynamic upset and requires no treatment.

Torsade de pointes VT

This is an uncommon but important form of VT, characterized by QRS complexes of changing amplitude that appear to twist around the isoelectric line and occur at a rate of 200–250 b.p.m. (Fig. 7.9). The arrhythmia is associated with prolongation of the Q–T interval, which can be due to drugs (Class I and Class III antiarrhythmic agents, phenothiazines, or tricyclic antidepressants) or congenital electrophysiologycal abnormalities. When this pattern of VT is seen with a long Q–T then:

1. Stop any precipitating drugs.
2. Correct hypokalaemia (which is frequently present) and give **magnesium** to patients treated with long-term diuretics.
3. Consider atrial or ventricular pacing at a rate of 90–120 b.p.m. If the cause of the long Q–T interval can be removed, pacing can be discontinued when the Q–T interval has returned to normal and the arrhythmia will be unlikely to recur.

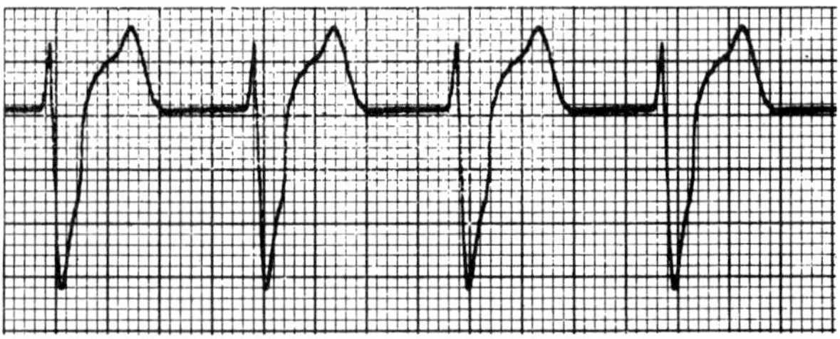

Fig. 7.8 Broad complex idioventricular rhythm with a rate of 35 b.p.m.

4. If the arrhythmia arises in a patient with congenital prolongation of the Q–T interval, IV beta-blockers may be effective.

Prognostic implications and long term drug therapy of ventricular arrhythmias

Patients who develop VF or sustained VT (ten or more successive beats) in the first 24 hours post-infarction, in the absence of heart failure, have a good prognosis after successful treatment of their arrhythmia. They do not require further investigation or long-term antiarrhythmic therapy.

Patients who develop sustained ventricular arrhythmias after the first 24 hours are at high risk of early death and should be considered for further investigation (including electrophysiological testing) to guide antiarrhythmic therapy, as this may improve their prognosis. If in doubt about the need for electrophysiological testing, consultation with the local invasive centre should clarify policy.

Although non-sustained ventricular arrhythmias occurring after the first 24 hours are a marker for a poor prognosis, they should not be treated if they are asymptotic since drug therapy with Class I agents in such patients was shown to have an adverse effect on prognosis in the CAST study.

Sustained ventricular arrhythmias that develop in a patient with overt heart failure post-infarction are associated with a poor prognosis. Even if initial resuscitation is successful, death often occurs in the next few weeks due to recurrent arrhythmia or progressive cardiac failure.

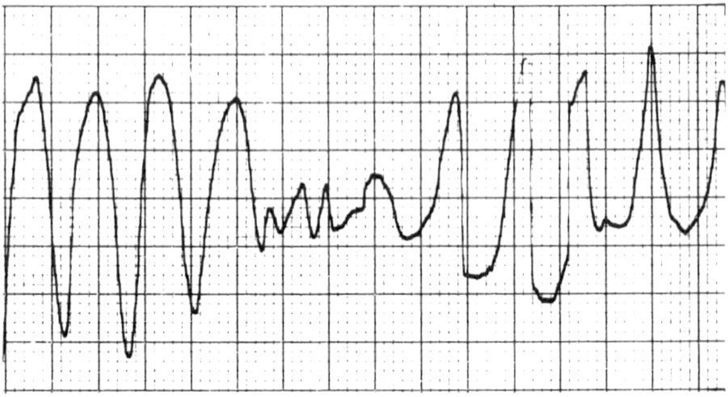

Fig. 7.9 *Torsades de pointes* VT, with QRS complexes of variable amplitude that appear to twist around the baseline

Bradyarrhythmias

Bradyarrhythmias are more common after inferior infarction, when their treatment is relatively simple and usually successful. If complete heat block develops in a patient with an inferior infarct, it is often preceded by progression through first-degree and Mobitz Type I block. The ventricular escape rhythm that develops is usually rapid enough to prevent haemodynamic collapse, and the patient's prognosis is good.

Bradyarrhythmias occurring in patients with anterior infarction are an ominous sign. They are usually associated with an extensive infarct and do not show an orderly progression to complete heart block with a stable escape rhythm. Treatment is more difficult and prognosis is poor even with appropriate treatment, due to associated extensive left ventricular damage.

Sinus bradycardia

This is common with inferior infarction and presents as sinus rhythm with normal AV conduction, and a rate below 60 b.p.m. (Fig. 7.10). If symptoms or ventricular

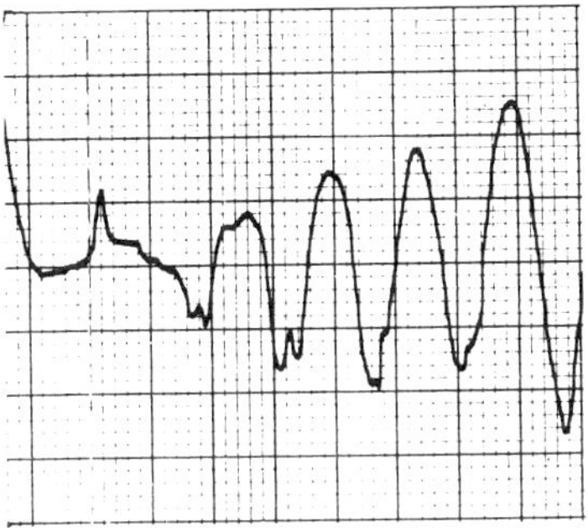

arrhythmias occur due to the slow heart rate, it should be treated with **atropine** 0.6 mg IV. If symptoms persist and atropine is unsuccessful, temporary pacing may be required.

First-degree heart block

This is common with inferior infarction. It is manifested as a prolongation of the P–R interval (Fig. 7.11), is associated with ischaemia of the atrioventricular node, and has a good prognosis. It requires no specific therapy, but should alert the doctor to the possibility of progression of the conduction disturbance to more serious higher degrees of block.

Mobitz Type I second-degree block (Wenckebach phenomenon)

This manifests itself as an increasing prolongation of the P–R interval with successive beats, until a QRS complex is dropped as AV conduction fails completely (Fig. 7.12). This type of second-degree block is usually associated with right coronary artery occlusion, inferior infarction and ischaemia at the level of the atrioventricular node, does not invariably progress to higher

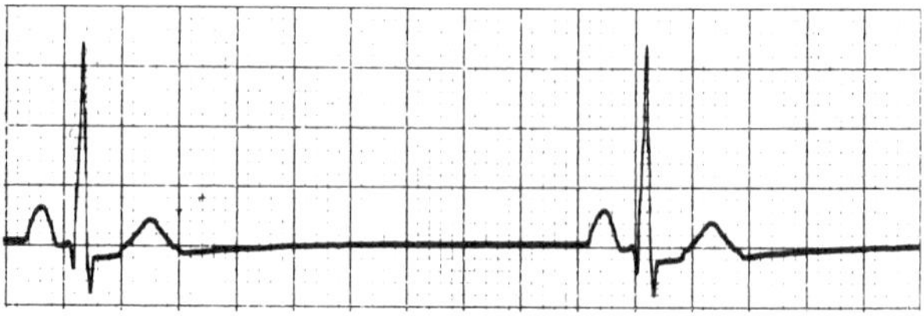

Fig. 7.10 Sinus bradycardia, with a rate of 30 b.p.m.

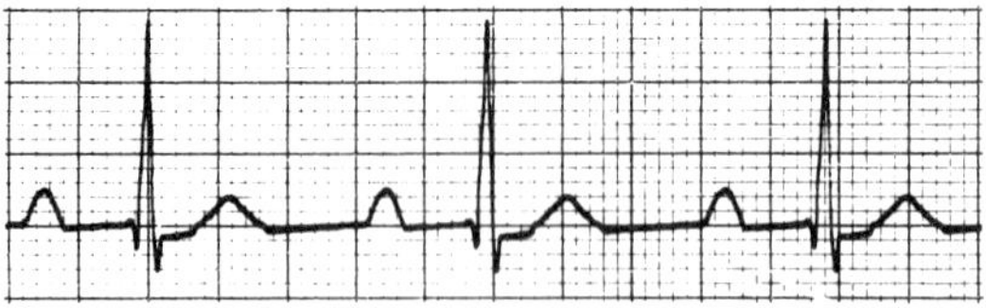

Fig. 7.11 Long P–R interval (0.3 s) in a patient with first-degree heart block (P–R interval normally 0.17–0.20 s)

degrees of block, and is not associated with a poor prognosis. No treatment is normally required, other than close observation for progression to higher degrees of block.

Mobitz Type II second-degree block

This manifests itself as sudden unpredictable failure of AV conduction resulting in a dropped beat, with no preceding change in the P–R interval (Fig. 7.13). It is usually associated with left anterior descending occlusion and extensive anterior infarction, is due to ischaemia in the conducting system below the level of the AV node, and frequently progresses to complete heart block with an unreliable slow escape rhythm. Its appearance in a patient with anterior infarction is an ominous sign, and these patients have a poor prognosis. Prophylactic

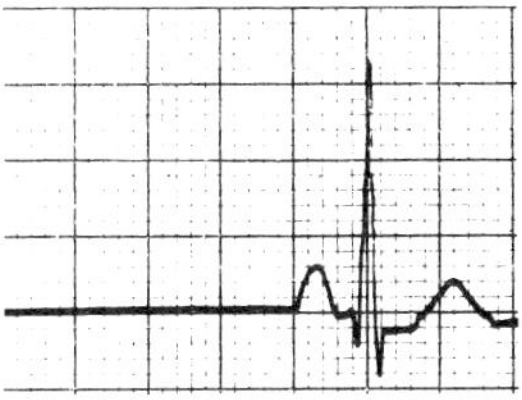

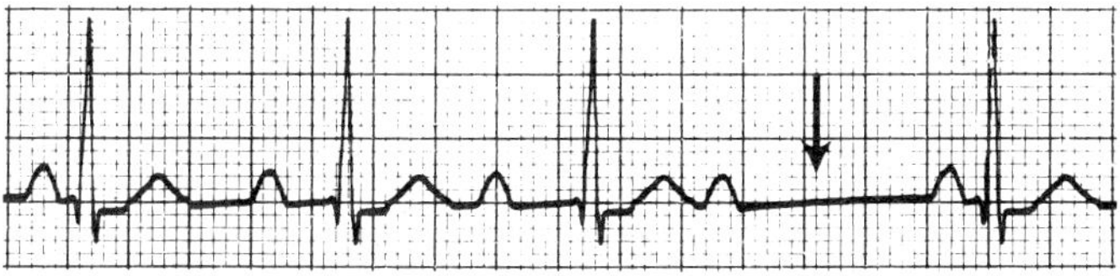

Fig. 7.12 Progressive prolongation of P–R interval from 0.2 s to 0.28 s to 0.33 s, followed by a dropped beat (arrowed) in a patient with Type I (Wenckebach) second-degree heart block

insertion of a pacing wire is recommended before complete heart block develops, as placement of a temporary pacemaker in a haemodynamically compromised patient with a slow, unreliable ventricular escape rhythm is likely to be difficult.

Third-degree (complete) heart block

This is manifest as complete dissociation between atrial and ventricular activity (Fig. 7.14). When it occurs in inferior infarction, prognosis is generally good. When complete heart block develops in anterior infarction, the in-hospital mortality is 80 per cent or more. It is associated with extensive infarction, may be preceded by the development of bundle branch block or Mobitz Type II second-degree block, and the escape rhythm that emerges is usually slow and unreliable and does not support the circulation.

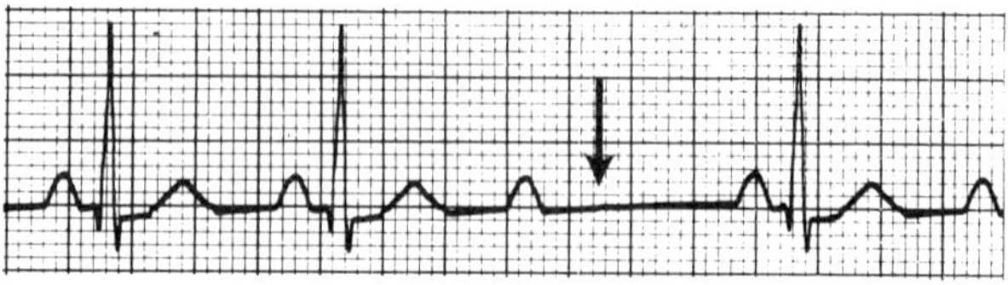

Fig. 7.13 Type II second-degree heart block. P–R interval of conducted beats is normal, with sudden failure of AV conduction resulting in an unexpected dropped beats (arrowed)

Bundle branch block

When left or right bundle branch block or the combination of bundle branch block and axis deviation (bifascicular block) develop in the course of an acute infarct, careful observation of the patient is indicated. These developments are usually seen in association with anterior infarction, and are markers of a poor prognosis. There is a significant risk of developing complete heart block in these patients, particularly if the P–R interval is also prolonged. Prophylactic temporary pacing is not mandatory (it has not been shown to improve prognosis), though opinion is divided on this point. Advice on local policy should be sought if a patient develops bifascicular block or bundle branch block in association with a prolonged P–R interval in the course of an infarct.

Drug therapy

Initial therapy for symptomatic bradyarrhythmias should be with **atropine** 0.6 mg IV. This can be repeated as necessary. If atropine fails, temporary support can be obtained with an **isoprenaline** infusion (as detailed in Chapter 18) prior to the insertion of a pacing wire.

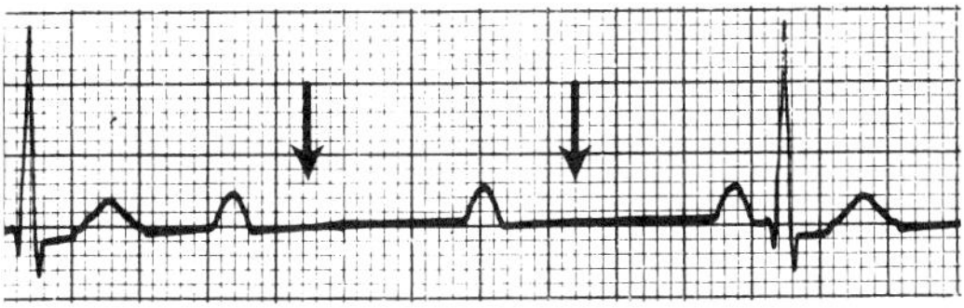

Temporary pacing post-infarction

Temporary cardiac pacing should not be used unless a definite indication is present. It is not an innocuous procedure and complications are common, particularly in inexperienced hands. Tachyarrhythmias induced during the procedure are particularly common post-infarction.

The following are indications for temporary pacing post-infarction:

- drug-resistant symptomatic sinus bradycardia, or symptomatic Type I second-degree heart block
- Type II second-degree block
- complete heart block in a patient with anterior infarction
- complete heart block in a patient with inferior infarction, if the ventricular rate falls below 50 b.p.m., the systolic BP falls below 100 mmHg, and rate-related symptoms or cardiac ischaemia develop
- prophylactic temporary pacemaker insertion for bundle branch block or bifascicular block, particularly if first-degree heart block is also present—this is a 'grey area' and advice on local policy should be sought
- overdrive pacing of tachyarrhythmias

The pacing wire should remain in place for 48 hours after the return of normal conduction following inferior

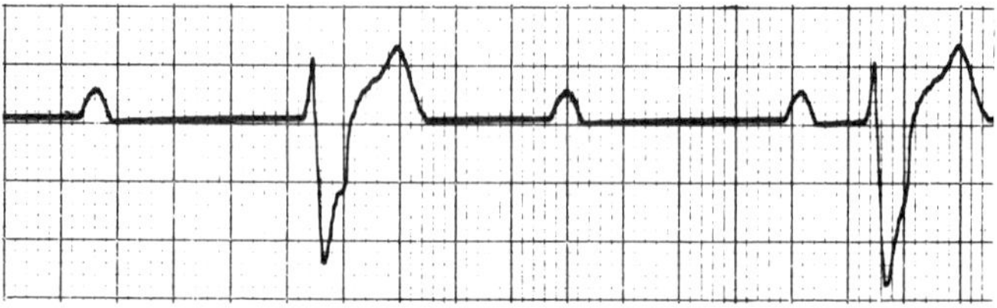

Fig. 7.14 Complete heart block showing slow ventricular escape rhythm (30 b.p.m.) with complete disassociation of independent atrial activity

infarction. In patients with anterior infarcts, it is preferable to leave the pacing wire *in situ* for 7 days following the return of normal conduction, as these patients are at risk of developing sudden complete heart block during this period. An algorithm to aid decision making in selecting patients who require temporary pacing is provided in Chapter 18.

The question of permanent pacing for survivors of acute myocardial infarction complicated by conduction defects remains controversial. Patients with ongoing symptomatic second- or third-degree heart block (i.e. that persists for more than 2 weeks post-infarction) require permanent pacing for symptomatic control. The remainder of patients are in a high-risk group for early mortality, but there is as yet no firm evidence that permanent pacing improves their prognosis.

Key points

- Arrhythmias are common after myocardial infarction. They do not all need aggressive treatment.
- Avoid negatively inotropic drugs in patients with a low blood pressure or severe impairment of ventricular function.

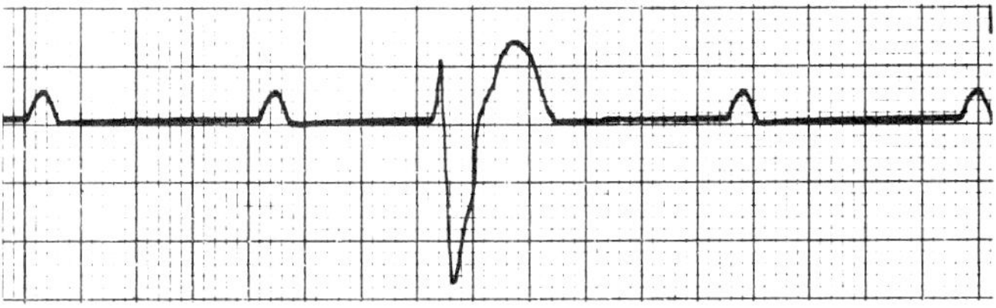

- Avoid polypharmacy and have a low threshold for DC cardioversion of any arrthymia in a patient who is deteriorating.
- Only use IV **verapamil** for patients with clear-cut paroxysmal supraventricular tachycardia and normal LV function.
- Do not routinely treat ventricular ectopics.
- In an older patient with a diagnosis of ischaemic heart disease or heart failure, there is a 90 per cent probability that a broad complex tachycardia will be VT.
- Late sustained ventricular arrhythmias are associated with a poor prognosis, and patients may benefit from drug therapy. Drug therapy for non-sustained or asymptomatic ventricular arrhythmis is not indicated.
- Conduction disturbances developing in a patient with anterior infarction are usually associated with an extensive infarct and a poor prognosis.

Further reading

Bhandari, A. K. and Rahimtoola, S. H. (1984). Indications for cardiac pacing in patients with bradyarrhythmias. *J.A.M.A.*, **252**, 1327–8.

Cowan, J. C. (1990). Management of supraventricular tachycardia. *Prescribers J.*, 95–101.

Dancy, M. and Ward, D. (1985). Diagnosis of ventricular tachycardia: a clinical algorithm. *Br. Med. J.*, **291,** 1036–8.

Garrat, C. J., Malcolm, A. D., Camm, A. J. (1992). Adenosine and cardiac arrhythmias. The preferred treatment for supraventricular tachycardia. *Br. Med. J.*, **305,** 3–4.

Griffith, J. J. and Camm, A. J. (1989). Broad QRS complex tachycardia. *Hosp. Update*, **15,** 531–41.

Griffith, M. J., Linker, N. J., Ward, D. E., and Camm, A. J. (1988). Adenosine in the diagnosis of broad complex tachycardia. *Lancet*, **i,** 672–5.

Hollander, G., Nadiminti, V., Lichstein, E., Greengart, A., and Sanders, M. (1983). Bundle branch block in acute myocardial infarction. *Am. Heart J.*, **105,** 738–43.

Huycke, E. C. and Sung, R. J. (1988). Management of supraventricular tachy-arrhythmias. *Chest*, **93,** 412–18.

Marchliuski, F. E. (1988). Treatment of sustained ventricular arrhythmias: which therapy to use. *Ann. Int. Med.*, **109,** 522–8.

Sapezynski, S. (1988). Update on antiarrhythmic drugs in emergency medicine. *Emerg. Med. Clin. N. Amer.*, **6,** 289–315.

Schutzenberger, W., Leisch, F., Kerschuner, K., Harringer, W., and Herbinger, W. (1989). Clinical efficacy of intravenous amiodarone in the short term treatment of recurrent sustained ventricular tachycardia and ventricular fibrillation. *Br. Heart J.*, **62,** 367–71.

The Cardiac Arrhythmia Suppression Trial (CAST) Investigators (1989). Preliminary report. Effect of encaindine and flecainide on mortality in a randomised trial of arrhythmia suppression after myocardial infarction. *N. Engl. J. Med.*, **321,** 406–12.

NOTES

NOTES

NOTES

NOTES

CHAPTER 8

Resuscitation

General principles

The guidelines set out in this chapter are based on the UK Resuscitation Council recommendations available in late 1991. In contrast to previous recommendations, **calcium** and **bicarbonate** are not used as first-line drugs (the evidence for their beneficial effect is marginal) while **adrenaline** is used much more liberally in prolonged resuscitation. The guidelines are presented in the form of a flow chart in Chapter 18.

Successful resuscitation from a ventricular arrhythmia in a patient with good left ventricular function should be the rule on the CCU. Asystole is more difficult to manage, and successful resuscitation is relatively rare.

When cardiac arrest occurs, basic life support should be instituted immediately. This consists of:

1. Confirm the presence of cardiac arrest—i.e. unconscious patient, absent respiration, and absent pulse in a major vessel.

2. Open the airway and start artificial respiration.
3. If no carotid pulse present after 2 ventilated breaths, start chest compression.
4. Obtain an ECG as soon as possible and treat as detailed below.

Remember to continue basic life support for 2 minutes after giving each drug to allow time for the drug to circulate. Do not interrupt basic life support for more than 10 seconds except for defibrillation.

In prolonged resuscitation, give 1 mg of adrenaline IV every 5 minutes, and 50 mmol of **sodium bicarbonate** (50 ml of 8.4% solution) after 10 minutes. Further administration of sodium bicarbonate should be guided by blood pH estimations if possible, aiming to keep pH between 7.3 and 7.5. If pH estimations are not available, give 25–50 ml every 10 minutes while resuscitation continues. **Calcium and bicarbonate should not be given through the same IV line**, or precipitation of the solution will occur and block the line. If IV access cannot be obtained in a patient suffering an unexpected cardiac arrest outside the CCU, drugs can be given in double doses via an endotracheal tube.

Ventricular fibrillation

This is recognized on the cardiac monitor by the presence of chaotic fibrillation waves due to wandering of cardiac electrical activity along continuously changing pathways (Fig. 8.1). A precordial thump administered to the lower sternum may revert the patient back to sinus rhythm. If this fails, proceed to DC cardioversion:

1. Initial treatment is a 200 J shock. Following each DC shock, 15 chest compressions should be given with appropriate ventilation, then a check of the rhythm and cardiac output made.

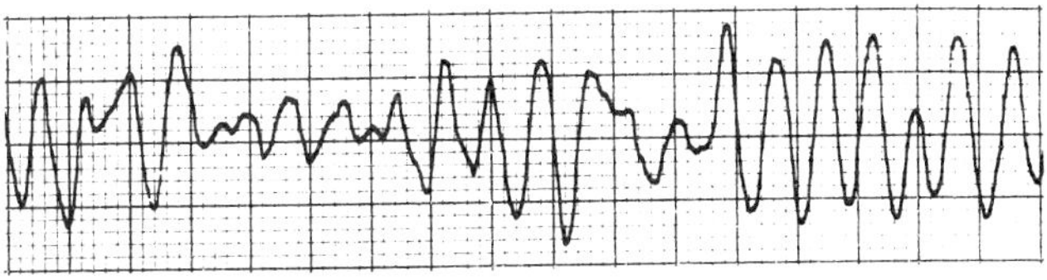

Fig. 8.1 Chaotic ventricular activity in ventricular fibrillation

2. If VF persists after the first shock, give a further 200 J shock, followed by a 360 J shock if necessary, continuing with basic life support between shocks.
3. If VF still persists, give 1 mg of **adrenaline** IV (1 ml of a 1:1000 solution), continue basic life support for 2 minutes to allow circulation, and repeat the 360 J shock.
4. If VF still persists, give **lignocaine** 100 mg IV, continue basic life support for 2 minutes, and repeat the 360 J shock.
5. If VF still persists, repeat the 360 J defibrillations trying
   (a) different paddle positions (ensuring good skin contact);
   (b) a different defibrillator;
   (c) other antiarrhythmic drugs.

The other drugs worth trying for refractory VF are **amiodarone** 300 mg IV or **bretylium** 5–10 mg/kg IV. (Remember that **bretylium** may take up to 20 minutes to work, so that resuscitation needs to continue for at least this length of time once this agent is given.)

After successful resuscitation from VF, consider a **lignocaine** or **amiodarone** infusion for the next 24 hours to suppress recurrent ventricular arrhythmias. The question of long-term antiarrhythmic therapy and the indications for electrophysiological testing are dealt with in Chapter 7.

Asystole

This is recognized by the total absence of any ventricular electrical activity on the cardiac monitor (Fig. 8.2). If any doubt exists in your mind that the trace is unreliable and that undetected VF may be present, initial

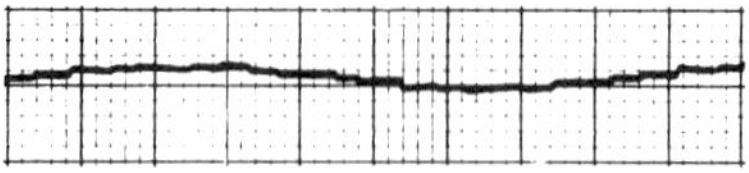

Fig. 8.2 Complete absence of cardiac electrical activity in asystole

treatment should be with three DC shocks (200 J, 200 J, then 360 J). If these are unsuccessful, proceed as detailed below for asystole:

1. Give **adrenaline** 1 mg IV, continue basic life support for 2 minutes, then check the rhythm and cardiac output.
2. If asystole persists, give **atropine** 2 mg IV, continue basic life support for 2 minutes, then check the rhythm and cardiac output.
3. If asystole persists beyond this point, the chance of successful resuscitation is very low. If any P-waves or QRS complexes are visible on the monitor, consider an **isoprenaline** infusion (5 mg of isoprenaline in 500 ml 5% **dextrose** infused at a rapid drip rate) or temporary pacing. If no electrical activity is present, the **adrenaline** and **atropine** can be repeated, but success is unlikely.

Electromechanical dissociation

This is recognized by the presence of ventricular complexes with an adequate rate on the monitor, with no cardiac output. The commonest cause of sudden, unexpected development of electromechanical dissociation on the CCU is rupture of the ventricular free wall. Successful resuscitation in this situation is very rare. Other important causes to consider are pneumothorax (has a central line been inserted recently?), pulmonary embolism (is the patient in a high risk group—i.e. bed-bound, low cardiac output state, or recent operation?), hypovolaemia (any obvious cause of major blood loss?), drug toxicity or allergy (any history of overdose recently?), and cardiac tamponade due to haemopericardium (any history of stab wound or aortic dissection?).

Having instituted basic life support and ascertained the presence of electromechanical dissociation:

1. Give **adrenaline** 1 mg IV.
2. Consider any specific therapy, i.e.:
   (a) pericardial aspiration for tamponade;
   (b) chest drain for pneumothorax;
   (c) volume replacement for hypovolaemia;
   (d) specific therapy for massive pulmonary embolism (Chapter 11).
3. **Calcium chloride** (10 ml of 10 per cent solution) should be given IV if hyperkalaemia, hypocalcaemia, or calcium antagonist overdose is a possibility.

If EMD persists and no treatable cause is apparent, **adrenaline** can be repeated, but successful resuscitation is unlikely.

Post-resuscitation care

Immediately after an arrest, check blood gases, electrolytes, and chest X-ray, and treat any problems, i.e.:

- In electromechanical dissociation, look for treatable causes before abandoning resuscitation.
- Careful post-resuscitation care is essential to maximize the chance of a full recovery.

Further reading

Camm, A. J. (1986). Asystole and electromechanical dissociation. *Br. Med. J.*, **292,** 1123–4.

Chamberlain, D. A. (1989). Advanced life support. *Br. Med. J.*, **229,** 446–8.

Chamberlain, D. (1986). Ventricular fibrillation. *Br. Med. J.*, **292,** 1068–70.

Evans, T. R. (ed.) (1990). *ABC of resuscitation*. British Medical Journal, London.

Marsden, A. K. (1989). Basic life support. *Br. Med. J.*, **299,** 442–5.

Skinner, D. V. (1991). Management of cardiac arrest. *Hosp. Update*, **17,** 416–20.

Yatsu, F. M. (1986). Cardiopulmonary–cerebral resuscitation. *New Engl. J. Med.*, **314,** 440–1.

NOTES

NOTES

NOTES

NOTES

CHAPTER 9

Left ventricular failure

General principles

Left ventricular failure is most common after extensive anterior infarction. Its occurrence is an adverse prognostic feature implying extensive ventricular damage with the loss of at least 20 per cent of the functional left ventricular myocardium.

The diagnosis of left ventricular failure should never be made on the basis of basal crepitations alone; these are often due to coexistent lung disease and diuretic therapy is inappropriate. The diagnosis is probable when two or more of the following occur:

- a prominent third heart sound is audible
- basal crackles, which do not clear with coughing, are present
- an extensive infarction has occurred
- radiological changes (upper lobe diversion, Kerley B lines, cardiomegaly, or alveolar oedema) are present

Most patients respond to first-line treatment with opiates, loop diuretics, and nitrates, as detailed below. Intravenous **diamorphine** is beneficial both for its sedative effects, and because it acts as a vasodilator. Intravenous loop diuretics have an immediate beneficial effect which precedes the onset of a diuresis, probably also due to vasodilatation. Reduction in cardiac preload and afterload with intravenous nitrates is beneficial where an adequate systolic blood pressure is present.

First-line therapy

Initial treatment for LVF should be:

1. Sit the patient up.
2. Give high-concentration oxygen therapy. If COPD is present, with its attendant risk of carbon dioxide retention, check blood gases and give controlled oxygen therapy as detailed in Chapter 18.
3. Give IV **diamorphine** 2.5 mg at 1 mg/min initially, plus **metoclopramide** 10 mg.
4. Give an IV loop diuretic—such as **frusemide** 40–80 mg IV.
5. If systolic BP is above 100 mmHg, give 600 μg of sublingual **GTN**, then commence an intravenous nitrate infusion as detailed in Chapter 18, increasing infusion rate to the maximum haemodynamically tolerated dose if left ventricular failure is severe.
6. Correct any precipitating arrhythmias or ongoing ischaemia that may be present and look for treatable mechanical problems (Chapter 12).

Second-line therapy

If initial treatment with **diamorphine**, **frusemide**, and **nitrates** fails, then:

1. Give further IV **frusemide** in double dose.
2. Give further **diamorphine** if the patient is still distressed.

3. Consider inotropic support with **dobutamine** and **dopamine** as detailed in Chapter 10 if systolic BP drops below 100 mmHg or urine output falls below 20 ml/hour for two consecutive hours

If the patient fails to respond to the above measures and there are no obvious reversible causes of left ventricular failure, such as ventricular septal rupture or acute mitral regurgitation, the outlook is poor, as failure to respond frequently implies severe ventricular impairment. Selective phosphodiesterase inhibition with **milrinone** or **enoximone** may be beneficial; however, it is not clear at present whether these drugs have any advantage above the two other inotropic agents mentioned above. These agents reduce pre- and afterload by dilating arteries and veins and are positively inotropic. They are contraindicated if the systolic BP is below 90 mmHg as they may cause a further fall in blood pressure. Advice from a senior colleague should be sought if the patient has severe left ventricular failure due to a mechanical complication of myocardial infarction, as positive pressure ventilation or mechanical support may improve the situation until corrective surgery can be undertaken.

Key points

- The occurrence of left ventricular failure is most common after an anterior infarction, and is indicative of extensive myocardial damage.
- Patients who do not respond to first-line therapy and have no easily reversible cause of their heart failure have a poor prognosis.
- Positive pressure ventilation or mechanical support is worth considering where a reversible cause of the left ventricular failure exists and drug therapy has failed.

Further reading

Flaherty, J. T., Reid, P. R., Kelly, D. T., Taylor, D. R., Weisfeldt, M. L., and Pitt, B. (1975). Intravenous nitroglycerin in acute myocardial infarction. *Circulation*, **51,** 132.

Glenville, B. (1991). Mechanical support for the failing heart. *Hosp. Update*, **17,** 89–97.

Vismara, L. A., Ledman, D. M., and Zelis, R. (1976). The effects of morphine on venous tone in patients with acute pulmonary oedema. *Circulation*, **54,** 335.

Wilson, J. R., Reichek, N., Donkman, W. S., and Goldberg, S. (1981). Effects of diuresis on the performance of the failing left ventricle in man. *Am. J. Med.*, **70,** 234.

NOTES

NOTES

NOTES

NOTES

CHAPTER 10

Cardiogenic shock

General principles

Patients with cardiogenic shock have severe acute heart failure secondary to a myocardial infarction. They have usually sustained a large infarct leading to loss of 40 per cent or more of their functioning left ventricular myocardium. This syndrome occurs in up to 15 per cent of patients who survive to reach hospital. Prognosis is poor, and the in-hospital mortality of established cardiogenic shock is around 90 per cent.

Some patients who present with apparent cardiogenic shock post-infarction will have circulatory compromise related to hypovolaemia, right ventricular infarction, or an acute mechanical complication of their infarct; the management and prognosis of these patients is different from that of patients with cardiogenic shock due to severe left ventricular impairment. Some physicians recommend pulmonary artery catheterization in all patients with a low output state following a myocardial infarction to exclude these treatable conditions. Facilities for pulmonary artery catheterization are not readily available in all hospitals, however. A diagnosis of cardiogenic shock due to severe impairment of left ventricular function can be confidently made in patients who fulfil the following criteria:

- recent extensive anterior infarction in a previously well patient
- recent myocardial infarction in a patient with known pre-existing impairment of LV function
- hypotension (systolic BP is usually well below 100 mmHg) and left ventricular failure
- clinical signs of a low output state—i.e. oliguria, impaired cerebration, peripheral hypoperfusion

In patients who fulfil these criteria, it is appropriate to institute supportive therapy without the aid of a pulmonary artery catheter.

If the patient has any of the following:

- hypotension without clinical or radiological evidence of pulmonary oedema
- hypotension in the presence of an inferior or posterior infarction in a patient with previously normal LV function
- hypotension and possible fluid depletion—i.e. a patient whose fluid intake is reduced or who has received aggressive diuretic therapy
- hypotension and any evidence of an acute mechanical complication—i.e. a new pansystolic murmur due to a VSD or severe acute mitral regurgitation,

then pulmonary artery catheterization is mandatory to clarify the diagnosis and guide therapy. The diagnosis and management of acute mechanical complications of myocardial infarction is dealt with in Chapter 12.

Remember that a low blood pressure alone does not mean that cardiogenic shock is present. Some patients with hypotension will be warm and well perfused with normal cerebration and a good urine output; these patients should be closely observed, and further investigations or treatment instituted only if other signs of cardiogenic shock or myocardial ischaemia due to hypotension develop.

Management of cardiogenic shock due to severe impairment of left ventricular function

1. Ensure that therapy with thrombolytic agents, correction of electrolyte disturbances, and treatment of arrhythmia or persistent myocardial ischaemia has been instituted where appropriate. Check potassium, glucose, and pH, and correct any associated metabolic abnormalities.
2. Insert a pulmonary artery catheter if doubt as to diagnosis exists. In cardiogenic shock due to severe LV impairment, pulmonary artery wedge pressure will be elevated.
3. Give **diamorphine** 2.5 mg IV at 1 mg/min plus **metoclopramide** 10 mg if the patient is distressed.
4. Give **oxygen** as detailed in Chapter 18 to relieve hypoxia.
5. Give **dobutamine** 2.5–20 μg/kg/min IV (see Chapter 18 for infusion chart). This is a beta-agonist and will increase cardiac output. Dobutamine can be given via a peripheral line.
6. Give **dopamine** 2.5–5.0 μg/kg/min IV via a central line if urine output is less than 20 ml/hour for two consecutive hours. Dopamine in this dose range is a selective renal artery vasodilator and may improve renal perfusion and thus urine output (see Chapter 18 for infusion chart).

7. If the blood pressure remains low despite the above measures, consider a **noradrenaline** infusion at 2–20 μg/min (see Chapter 18 for infusion chart).
8. Treat coexistent LVF with IV loop diuretics. If the above inotropic support increases systolic BP to 100 mmHg, intravenous **nitrates** can be carefully introduced to offload the left ventricle using the regimes detailed in Chapter 18, commencing at a low dose initially and titrating the dose carefully upwards if blood pressure is maintained.

If the patient fails to respond to the above measures, and a low left atrial pressure or mechanical complication have been excluded, then the outlook is very poor. Intra-aortic balloon pumping is unlikely to be effective if there is no treatable mechanical complication underlying the cardiogenic shock. The use of selective phosphodiesterase inhibitors such as **milrinone** or **enoximone** can be considered if the systolic BP is at least 90 mmHg, to provide additional inotropic support and pre- and afterload reduction. (These agents, however, tend to cause hypotension and the patient must be carefully monitored for this adverse effect.) In selected patients with cardiogenic shock due to acute myocardial infarction emergency PTCA may improve the outlook; following the institution of initial resuscitation all patients with cardiogenic shock should be discussed with a senior colleague to ensure that further investigations are not indicated.

Right ventricular infarction

This syndrome occurs in 5–10 per cent of patients with inferior or posterior infarction, and can produce a clinical picture of cardiogenic shock. Unlike cardiogenic shock due to extensive left ventricular infarction, prognosis with appropriate treatment is relatively good, with survival rates of 60 per cent quoted in some studies.

Right ventricular infarction should be suspected where:

- hypotension exists in a patient with an inferior or posterior infarct
- hypotension exists with a clear chest on auscultation and clear lung fields on a chest X-ray
- raised JVP, pulsus paradoxus, or Kussmaul's sign is present

If right ventricular infarction is suspected:

1. Record a V4R lead. 1 mm S–T elevation in this lead is a relatively sensitive and specific sign of right ventricular infarction
2. Insert a pulmonary artery pressure line and look for the characteristic pressure profile. The pulmonary artery wedge pressure will be low or normal, but the right ventricular diastolic pressure and right atrial pressure will be elevated.

When a diagnosis of right ventricular infarction has been made, treat with intravenous fluids to raise the PA wedge pressure to 16–18 mmHg. If hypotension persists after this, add inotropic support. Diuretics should be avoided as they will decrease the circulating volume and exacerbate the situation.

Hypovolaemia

If hypovolaemia is a possibility, insert a pulmonary artery pressure line and look for the characteristic pressure profile. Pulmonary artery wedge pressure, right ventricular and right atrial pressure will all be low. Treatment is with intravenous fluids to raise the wedge pressure to 15–18 mmHg. If hypotension persists despite adequate volume replacement, add inotropic therapy.

Key points

- Cardiogenic shock can occur in up to 15 per cent of patients who survive to reach hospital.
- In-hospital mortality is high (90 per cent).
- First-line therapy is with **diamorphine**, **oxygen**, and **inotropic** agents.
- Right ventricular infarction or hypovolaemia should be excluded by pulmonary artery catheterization if any diagnostic doubt exists.
- Where a possible treatable cause exists (ventricular septal rupture, or severe acute mitral regurgitation) then consideration of mechanical support as a bridge to surgery is appropriate.

Further reading

Balakumaran, K. and Hugenholtz, P. G. (1986). Cardiogenic shock: current concepts in management. *Drugs*, **32,** 372–82.

Caplin, J. L. (1989). Acute right ventricular infarction. *Br. Med. J.*, **299,** 69–70.

Rabinowitz, B., Tamari, I., Elazar, E., and Neufeld, H. N. (1982). Intravenous isosorbide dinitrate in patients with refractory pump failure and acute myocardial infarction. *Circulation*, **65,** 771–8.

Rackley, C. E., Russel, R. O., Mantle, J. A., and Moraski, R. E. (1975). Cardiogenic shock: recognition and management. *Cardiovasc. Clinic*, **7,** 251–8.

NOTES

NOTES

NOTES

NOTES

CHAPTER 11

Anticoagulants and antiplatelet agents

Prophylactic subcutaneous heparin

Acute myocardial ischaemic syndromes are associated with an increased risk of deep venous thrombosis and pulmonary embolism. Prophylactic low dose subcutaneous **heparin** 5000 U t.d.s. has been shown to decrease the incidence of deep venous thrombosis and pulmonary embolism, and should be used in all patients during periods of enforced bed rest.

This regime is not associated with an increased risk of bleeding complications, and does not require any haematological monitoring. The **heparin** should be discontinued when the patient is fully ambulant.

Aspirin in acute ischaemic syndromes

Aspirin has been shown to be of unequivocal benefit in reducing mortality in acute myocardial infarction and unstable angina. Controversy exists as to the optimal dose regime. Those which are known to be effective are:

- 160 mg daily in acute infarction
- 325 mg daily in unstable angina

The first dose should be given as soon as the diagnosis is made, and should be sucked or chewed to facilitate rapid absorption.

Relative contraindications to the use of **aspirin** are a history of allergy, active peptic ulcer disease, or recent major trauma. A history of gastrointestinal intolerance is not an absolute contraindication, since this can be minimized by the use of low-dose maintenance regimes.

Aspirin should be continued for an indefinite period at a dose of 160 mg or 325 mg daily. If troublesome dyspepsia occurs, 75 mg daily is a low-dose alternative for which there is some evidence of efficacy. There is at present no unequivocal evidence to support the use of aspirin as primary prophylaxis in patients considered to be at high risk for a cardiac ischaemic event.

Anticoagulation and thrombolysis

Anticoagulation in patients who have received thrombolytic therapy remains a controversial area. The combination of **streptokinase** and **aspirin** has been shown to be effective without addition of anticoagulation in both the GISSI and ISIS studies. Since additional anticoagulation increases the risk of bleeding complications without producing a significant improvement in outcome, we do not recommend its routine use after streptokinase.

The thrombolytic state induced by the use of **tissue plasminogen activator** is relatively short-lived compared with streptokinase, and there is therefore a higher risk of early vessel reocclusion. There is some evidence that this risk can be reduced by the use of anticoagulation, and we recommend commencing a **heparin** infusion at 1000 U per hour after the cessation of the tissue plasminogen activator infusion. The infusion should be continued for 48 hours. The APIT should be checked after six hours and the infusion titrated to maintain this at 2–3 times the control value. Overanticoagulation should be carefully avoided as this may increase the risk of stroke.

At the present time there are no published data to support the long-term use of oral anticoagulants following thrombolysis, and we do not recommend their routine use.

Anticoagulation and unstable angina

There is now evidence that **heparin** improves the outlook in patients with unstable angina. We currently use a heparin infusion (with the infusion rate titrated to maintain the APTT at 2–3 times the control value) when first-line therapy with **beta-blockers**, **nitrates**, and **aspirin** has failed. Evidence supporting the early use of heparin in these patients is accumulating, and some physicians may advocate its early use as first-line therapy. You should obtain guidelines on local policy from a senior colleague at the start of a CCU appointment.

Anticoagulation and thromboembolism

Patients with a deep venous thrombosis or pulmonary embolism (see Chapter 12) should be anticoagulated with **heparin** using the APTT to control the infusion rate. **Heparin** should be continued for at least 5 days and followed by oral anticoagulants for 3–6 months.

The following patients are at increased risk of developing thromboembolic complications after an acute myocardial infarction:

- Patients with large anterior or apical infarcts, since 30 per cent will develop mural thrombus, which may give rise to systemic emboli.
- Obese immobile patients, particularly if they have a complicated in-hospital course associated with prolonged bed rest, are at high risk of deep venous thrombosis and pulmonary embolism.
- Patients who survive their infarct but are left with severe impairment of left ventricular function are likely to develop thrombus in a dilated dyskinetic left ventricle or fibrillating left atrium, and are therefore at high risk of systemic embolization.
- Patients with a past history of thromboembolism are likely to develop a recurrent episode.

Patients in these four categories should be considered for immediate anticoagulation with a **heparin** infusion, as this may reduce the incidence of thromboembolic complications. Not all physicians advocate full systemic anticoagulation in these patients, and you should consult a senior colleague for advice on local policy. A pericardial friction rub is common following an infarct, and is not an absolute contraindication to anticoagulation, but the increased risk of haemopericardium should be borne in mind.

Most physicians would advocate oral anticoagulation for 3–6 months in patients who have had a DVT or pulmonary embolism. Long term anticoagulation is indicated in patients who have had a systemic embolism, who have mural thrombus detectable by echocardiography, or who survive their infarct with severe impairment of left ventricular function.

Key points

- All patients are at risk of thromboembolism during the time their mobility is restricted, and should receive a low dose of subcutaneous **heparin**.
- Aspirin is of proven efficacy and should be given to all patients with an acute myocardial infarction or unstable angina.
- Heparin is not routinely indicated after thrombolysis with **streptokinase**. All patients receiving **tissue plasminogen activator** should be treated with intravenous heparin.
- Heparin is presently used in unstable angina if the patient fails to settle with standard first-line drug therapy.
- Patients at high risk of thromboembolic complications should be considered for short-term anticoagulation. Patients who have had an embolic episode, or who have severe impairment of left ventricular function, should be considered for long-term oral anticoagulants.

Further reading

Gruppo Italiano per Lo studio Della Streptochinas; nell infarto Miocardico (GISSI) (1986). Effectiveness of intravenous thrombolytic treatment in acute myocardial infarction. *Lancet*, **i**, 397–402.

ISIS-2 (Second International Study of Infarct Survival) Collaborative Group (1988). Randomised trial of intravenous streptokinase, oral aspirin, both, or neither among 17,187 cases of suspected acute myocardial infarction. *Lancet*, **ii**, 349–60.

Kakkar, V. V. and Adams, P. C. (1986). Preventative and therapeutic approach to venous thromboembolic disease and pulmonary embolism. Can death from pulmonary embolism be prevented. *J. Am. Coll. Cardiol.*, **8,** 146–58B.

Neri Serneri, G. G., Gensini, G. F., Poggesi, L., *et al.* (1990). Effect of heparin, aspirin or alteplase in reduction of myocardial ischaemia in refractory unstable angina. *Lancet*, **335,** 615–18.

Orme, M. (1988). Aspirin all round. *Br. Med. J.*, **296,** 307–8.

Theroux, P., Ouimet, H., McCans, J., *et al.* (1988). Aspirin, heparin or both to treat acute unstable angina. *N. Engl. J. Med.*, **319,** 1105–11.

Thromboembolic Risk Factors (THRIFT) Consensus Group (1992). Risk of and prophylaxis for venous thromboembolism in hospital patients. *Br. Med. J.*, **305,** 567–74.

Wray, R., Maurer, B., and Shillingford, J. (1973). Prophylactic anticoagulant therapy in the prevention of calf vein thrombosis after myocardial infarction. *New Eng. J. Med.*, **288,** 815–17.

NOTES

NOTES

NOTES

NOTES

CHAPTER 12

Complications occurring 1–14 days post-myocardial infarction

Pericarditis

Acute pericarditis occurs in 20 per cent of patients, usually on the second or third day post-infarction. It is usually associated with transmural infarcts, and presents as pericardial pain associated with a friction rub. It is benign and self-limiting, and should be treated with simple analgesics such as **Dihydrocodeine** 30 mg four to six hourly, since non-steroidal agents such as Indomethacin may encourage infarct expansion. Pericarditis occurring later (usually 2–6 weeks post-infarction) may be associated with pleurisy and pleural effusion (Dressler's syndrome) and should be treated initially with non-steroidal anti-inflammatory drugs.

If this treatment fails, oral steroids should be given. Dressler's syndrome has been described in 4 per cent of post-infarct patients, with recent reports suggesting a much lower incidence.

Rupture of the ventricular free wall

This occurs in approximately 8 per cent of all transmural infarcts between the first and tenth day and accounts for 10 per cent of all in-hospital deaths. It is more common in older patients, women, hypertensives, and those sustaining their first infarction. It presents as sudden cardiovascular collapse with electromechanical dissociation and resuscitation is rarely successful. In both ISIS 2 and GISSI I, there was an excess of deaths in the early post-infarct hours in patients given thrombolysis, due to an increased incidence of cardiac rupture. Since intravenous beta-blockers are known to reduce the incidence of this complication, there is a rational basis for using these agents in combination to reduce this risk.

Rupture of the intraventricular septum

This occurs in about 1 per cent of acute infarcts, usually between days seven to ten. It is more common after anterior than inferior infarction and presents with the sudden onset of heart failure and the development of a new pansystolic murmur at the lower left sternal edge. The diagnosis can be confirmed by colour flow Doppler echocardiography or right heart catheterization (see the following section on papillary muscle dysfunction and rupture for details). The majority of patients are severely haemodynamically compromised, and mortality is high unless the defect can be closed surgically in the early stages. In general, there is no benefit in deferring surgery. In the minority of patients who are haemodynamically stable despite the presence of an acute ventricular septal rupture, deferring surgery for 6 weeks may reduce the operative risk and increase the chances of successful surgical repair. Once the diagnosis has been made and full supportive therapy started, the patient should be transferred to a cardiothoracic unit for consideration of surgery. If the patient's clinical state cannot be stabilized with drug therapy, it is worth considering intra-aortic balloon pumping until surgical correction can be attempted.

Papillary muscle dysfunction

A minor degree of mitral regurgitation is common following a myocardial infarction, due to dilatation of the left ventricle. These patients have a quiet murmur and are not haemodynamically compromised. Severe acute mitral regurgitation due to complete or partial rupture of a papillary muscle is a rare complication of acute MI (0.5–1.0 per cent). It commonly involves the posterior papillary muscle as part of an inferior infarct. The patient presents with sudden onset of severe left ventricular failure on days two to ten with a new, loud apical pansystolic murmur. It can be difficult to differentiate this from an acute ventricular septal rupture, but echocardiography and right heart catheterization will help to make the distinction. The important clinical and investigational features are as follows:

- Murmur tends to be loudest at the apex in acute mitral regurgitation, and loudest at the lower left sternal edge with a ventricular septal rupture.
- Colour flow Doppler may show a regurgitant jet crossing the septum in a ventricular septal rupture. In acute mitral regurgitation, prominent regurgitant flow will be detected through the mitral valve, and

the two-dimensional image may show ruptured papillary muscles or a flail mitral leaflet.

- In a ventricular septal rupture there will be a 'step up' in oxygen saturation in blood samples from the right ventricle and pulmonary artery compared with those from the right atrium, due to oxygenated blood crossing the septum from left to right ventricle.
- In acute mitral regurgitation, there will be no 'step up' in oxygen saturation. Prominent V-waves may be visible in the PA wedge pressure tracings.

Mortality is high in these patients (50 per cent within 24 hours, 90 per cent at 2 weeks), who usually have severe heart failure refractory to medical therapy. Surgical mortality is 25 per cent but operative correction of the defect offers the only hope of survival. As soon as the diagnosis is made and maximal supportive therapy has been instituted, the patient should be referred to a cardiac surgeon. Intra-aortic balloon pumping may be necessary to stabilize the clinical condition. Intra-aortic balloon pumps require experience and skill to insert and use. The balloon in the aorta inflates during diastole, increasing aortic diastolic pressure and improving coronary perfusion. The balloon deflates during systole, reducing afterload and thereby improving left ventricular function and cardiac output.

Ventricular aneurysm

Ventricular aneurysms develop in 10–15 per cent of survivors post-infarct. They are four times more common after transmural anterior than inferior infarction. The development of an aneurysm is frequently asymptomatic. A dyskinetic apical impulse or persistent S–T elevation on the ECG should raise the possibility of a left ventricular aneurysm. Surgical intervention to resect an aneurysm is only indicated if the complications of tachyarrhythmias, intractable heart failure, or embolization develop. If facilities permit, all infarct patients (particularly those with a transmural anterior infarct) should be screened for left ventricular aneurysm and mural thrombus with two-dimensional echocardiography prior to discharge. Patients who have a mural thrombus should be anticoagulated.

Infarct extension

This is suggested by recurrence of ischaemic pain at rest with associated ECG changes of ischaemia or infarction and a further rise in cardiac enzymes. It occurs in up to 10 per cent of patients during the first week and is associated with a poor prognosis. These patients should be considered for urgent angiography with a view to revascularization, as this may improve their outlook.

Deep venous thrombosis

DVT is common in hospital patients. Half of all DVT's are clinically silent. Many patients who have an episode of pulmonary embolism do not have a clinically obvious DVT. Thrombus is present in the pelvic veins in most of these patients. Patients at high risk of DVT are those with:

- heart failure following an infarct
- recent surgery
- malignant disease
- immobility
- previous history of thromboembolism
- obesity

The usual presentation is with a unilateral painful,

swollen, warm leg (remember that there are other possible causes of a painful swollen leg). Treatment is with IV **heparin** (5000 U IV as a loading bolus over 5 minutes, followed by an infusion of 1000 U/hour for 6 hours. The APTT should then be checked and the subsequent infusion rate adjusted to maintain this at 2–3 times the control value). The diagnosis should be confirmed by venography if possible. Oral anticoagulants should be commenced before the heparin is discontinued, and continued for 3–6 months.

There is debate about the value of anticoagulating DVTs confined to the calf and you should seek advice on local policy. Low-dose subcutaneous heparin is known to reduce the incidence of DVT post-infarction, and all patients should receive this regime prophylactically until ambulant.

Pulmonary embolism

Pulmonary embolization is not uncommon in patients treated on the CCU. Patients who have a complicated course following an infarct are at particular risk. These patients often have a low-output state and a prolonged period of bed rest, both of which promote the development of venous thrombosis with subsequent pulmonary embolism. The presence of other risk factors for DVT increase this risk still further.

A large embolus presents with sudden onset of acute dyspnoea, which may be accompanied by chest pain. Cardiovascular collapse develops due to the obstruction to outflow from the right ventricle, and the patient develops one or more of the following clinical signs:

- hypotension
- tachycardia
- cyanosis

- tachypnoea
- sweating
- wheeze
- raised JVP

Remember that many DVT's are clinically silent, and the absence of a swollen leg does not rule out the diagnosis of pulmonary embolism.

A small- to medium-sized embolism commonly present in a less dramatic fashion with dyspnoea, unilateral pleuritic chest pain, and haemoptysis due to pulmonary infarction.

The ECG is normal in the majority of patients with acute pulmonary embolism, apart from the presence of a sinus tachycardia. The well known ECG signs of S1, Q3, T3, right bundle branch block, right axis deviation, and right ventricular strain are often transient, and only seen with a large embolism. Atrial fibrillation may be precipitated by a pulmonary embolism. Similarly, the chest X-ray is usually normal with an acute large embolus. With small to medium emboli, wedge-shaped shadows or pleural effusions due to pulmonary infarction may be seen. Arterial blood gases are usually abnormal, showing hypoxia with a low PCO_2 due to the combination of VQ mismatching and hyperventilation.

The diagnosis of acute pulmonary embolism is usually made on clinical grounds, in a patient who is at high risk and develops clinical features of pulmonary embolism with hypoxia and a low PCO_2. Treatment should be instituted immediately the diagnosis is made. Confirmation of the diagnosis by isotope lung scanning or pulmonary angiography should not delay treatment.

Treatment of acute pulmonary embolism consists of:

1. Give IV **diamorphine** 2.5 mg at 1 mg/minute plus **metoclopramide** 10 mg to relieve distress.
2. Give high concentration **oxygen** to relieve hypoxia.

3. Anticoagulate with IV **heparin** using the regime described for DVT to prevent the development of further thrombus.
4. Support the circulation with inotropic agents.

If hypotension persists despite the above measures, mortality without further aggressive therapy is likely to be high. These patients require emergency pulmonary angiography. Pulmonary angiography allows visualization of the embolus and confirmation of the diagnosis. The catheter can then be used to administer thrombolysis using 250 000 U of **streptokinase** in 100 ml of 5% **dextrose** over 30 minutes, followed by 100 000 U per hour for 24 hours. Attempts to mechanically disrupt the clot with the angiographic catheter, or surgical embolectomy may be undertaken if there is local expertise in these techniques.

Following an acute pulmonary embolism, heparin should be continued for at least 5 days. Oral anticoagulants should be continued for 6 months. Patients who survive an episode of acute pulmonary embolism usually make a full recovery.

Key points

- Pericarditis is common and easily treated.
- Acute ventricular septal and papillary muscle rupture usually present as sudden-onset severe heart failure associated with a new pansystolic murmur. Echocardiography and pulmonary artery catheterization are invaluable in differentiating these conditions, both of which may be amenable to surgical correction.
- Evidence of continuing ischaemia or infarct exten-

sion should lead to early angiography with a view to revascularization.

- Many patients with a large pulmonary embolism have no diagnostic abnormalities on their chest X-ray or ECG.
- If a patient with a pulmonary embolism remains compromised despite oxygen, analgesia, and inotropes, consider urgent pulmonary angiography with a view to thrombolysis or surgical removal of the clot.

Further reading

Bates, R. J., Beutler, S., Resnekov, L., and Anagostopoulos, C. E. (1977). Cardiac rupture—challenge in diagnosis and management. *Am. J. Cardiol.*, **40**, 429–37.

Dressler, W. (1959). The post myocardial infarction syndrome: a report of forty four cases. *Arch. Intern. Med.*, **103**, 28–42.

Fox, A. L., Glassman, E., and Isom, O. W. (1979). Surgically remediable complications of acute myocardial infarction. *Prog. Cardiovasc. Dis.*, **21**, 461–84.

Hilton, C. J. (1991). Surgical management of complications of myocardial infarction. *Hosp. Update*, **17**, 197–209.

Hopkins, N. F. G. and Wolfe, J. N. H. (1991). Thrombosis and pulmonary embolism. *Br. Med. J.*, **303**, 1260–2.

Lancet, (1988). Management of venous thromboembolism. *Lancet*, **i**, 275–7.

Marmor, A., Sobel, B. E., and Roberts, R. (1981). Factors presaging early recurrent myocardial infarction (extension). *Am. J. Cardiol.*, **48**, 603–10.

Nishimura, R. A., Schaff, H. V., Shubb, C., Gersh, B. J., Edwards, W. D., and Tajik, A. J. (1983). Papillary muscle rupture complicating acute myocardial infarction: analysis of 17 patients. *Am. J. Cardiol.*, **51**, 373–7.

Seth, A., Perry, R. A., and Shiu, M. F. (1988). Management of unstable angina soon after myocardial infarction. *Br. J. Hosp. Med.*, **39**, 400–8.

Vlodaver, Z. and Edwards, J. E. (1985). Rupture of ventricular septum or papillary muscle complicating myocardial infarction. *Circulation*, **55**, 815–22.

NOTES

NOTES

NOTES

CHAPTER 13

Unstable angina

Background

Unstable angina is a common clinical syndrome, which often precedes acute infarction. A useful working definition is angina pectoris that is increasing in frequency, duration, and intensity in the absence of myocardial necrosis. The patient with unstable angina is optimally managed on the CCU at least during the initial unstable phase. A cardiac enzyme and ECG series following admission with unstable angina is necessary to exclude infarction.

A 12-lead ECG should be recorded during any episodes of chest pain in order to detect S–T segment and T-wave changes. If the resting ECG is normal and fails to show any changes during pain, it is less likely that the pain is cardiac in origin, and further investigation to define the cause of the pain should be pursued. S–T-segment depression and T-wave inversion are the commonest ECG changes, their distribution indicating which coronary artery is implicated. S–T elevation, T-wave peaking, and pseudo-normalization of T-waves during pain are less common and occur with variant angina, due to predominant coronary vasoconstriction.

First-line therapy

1. Admit to the CCU for bed rest and treatment.
2. Insert an IV line.
3. Give soluble **aspirin** 325 mg. This dose is known to be effective in reducing mortality by 50 per cent. A dose of 75 mg daily has recently been shown to be effective in unstable angina, although onset of full antiplatelet activity may be delayed for 48 hours with this low-dose regime, and we prefer to commence therapy initially with 325 mg.
4. Commence beta-blockers on admission using a shorter-acting agent, such as **metoprolol** 50 mg b.d.. This reduces myocardial oxygen demand and prevents undue tachycardia induced by vasodilators. Beta-blockers are effective agents in controlling unstable angina when predominant coronary spasm is excluded.
5. Start IV **nitrates** commencing at a low dose and increasing as necessary if further episodes of pain occur (see Chapter 18 for infusion charts).
6. If **aspirin**, **nitrates**, and beta-blockers fail to control the unstable angina, add a calcium antagonist. **Nifedipine** 10–20 mg t.d.s. can safely be used with beta-blockers. If variant angina is suspected, beta-blockers should be avoided and calcium antagonists started at an early stage.

Second-line therapy

The majority of patients will settle on combination therapy as outlined above and can then be further investigated electively. A small proportion of patients will remain unstable. Second-line therapy for these patients consists of full systemic anticoagulation with a **heparin** infusion using the APTT to control infusion rate (2–3 times the control value) continued until the acute episode has stabilized. Thrombolysis has not yet been shown to have any benefit in unstable angina. It should only be used where episodes of pain become prolonged and definite evidence of infarction (in the form of S–T elevation) develops. If pain is persistent, ECGs should be repeated at 20-minute intervals to detect evidence of evolving infarction.

All unstable patients who fail to settle on first-line therapy should be considered for early angiography with a view to revascularization. If facilities for coronary angiography are not available on site, contact should be made with the local invasive centre to discuss transfer for further investigation. The evidence that **heparin** is beneficial at an early stage is accumulating, and some physicians may prefer to anticoagulate patients with unstable angina as first-line therapy. You should clarify this point with a senior colleague at the start of your CCU post.

Subsequent management

Ideally, all patients suitable for revascularization procedures should undergo exercise testing prior to hospital discharge. Exercise tests positive at low workload identify a group of patients who are likely to require intervention with surgery or angioplasty. In a non-invasive centre, the results of the exercise test can provide a basis for selecting those patients to refer for early angiography at a local invasive unit.

Key points

- All patients with unstable angina should be admitted to the CCU.

- An ECG should be recorded during an episode of pain. If no ECG changes are present during pain an alternative diagnosis is more likely.
- If an unusual ECG pattern is shown during pain, with S–T elevation, T-wave peaking, and pseudo-normalization of T-waves, the diagnosis of variant angina is a possibility, and beta-blockers should be avoided.
- Thrombolysis is not indicated unless clear-cut evidence of developing infarction is present.

Further reading

Cairns, J. A., Gent, M., Singer, J., *et al.* (1985). Aspirin, sulphinpyrazone or both in unstable angina—results of a Canadian multicentre trial. *N. Engl. J. Med.*, **313,** 1369–75.

de Zwaan, C., Bar, F. W., Janssen, J. H. A., DeSwart, H. B., Vermeer, F., and Wellens, H. J. J. (1988). Effects of thrombolytic therapy in unstable angina: clinical and angiographic results. *J. Am. Coll. Cardiol.*, **12,** 301–9.

Falk, E. (1985). Unstable angina with fatal outcome dynamic coronary thrombosis leading to infarction and/or sudden death: autopsy evidence of recurrent mural thrombosis with peripheral embolisation culminating in total vascular occlusion. *Circulation*, **71,** 699–708.

Hargreaves, A. D. and Boon, N. A. (1991). Management of unstable angina. *Hosp. Update*, 98–111.

Holmes, Jr, D. R., Hartzler, G. O., Smith, H. C., and Fuster, V. (1981). Coronary artery thrombosis in patients with unstable angina. *Br. Heart J.*, **45,** 411–16.

Lewis, Jr, H. D., Davis, J. W., Archibald, D. G., *et al.* (1983). Protective effects of aspirin against acute myocardial infarction and death in men with unstable angina: results of a Veterans Administration Cooperative Study. *N. Engl. J. Med.*, **309,** 396–403.

Mulcahy, R., Daly, L., Graham, I., *et al.* (1981). Unstable angina: natural history and determinants of prognosis. *Am. J. Cardiol.*, **48,** 525–8.

Neri Serneri, G. G., Gensini, G. F., Poggesi, L., *et al.* (1990). Effect of heparin, aspirin or alteplase in reduction of myocardial ischaemia in refractory unstable angina. *Lancet*, **335,** 615–18.

Telford, A. M. and Wilson, C. (1981). Trial of heparin versus atenolol in prevention of myocardial infarction in intermediate coronary syndrome. *Lancet*, **i,** 1225–8.

Theroux, P., Ouimet, H., McCans, J., *et al.* (1988). Aspirin, heparin, or both to treat acute unstable angina. *N. Engl. J. Med.*, **319,** 1105–11.

NOTES

NOTES

NOTES

CHAPTER 14

The recovery phase

Mobilization, discharge, and follow-up

Patients should be kept on bed rest for the first 24 hours post-infarction. If their clinical course is uncomplicated they can then be transferred from the CCU and mobilized gently on the ward. Patients should be discharged 7–10 days post-infarction if they have made an uncomplicated recovery. Patients with complicated infarcts will require a longer time on the CCU and a more prolonged period of hospitalization. Mobilization should be more cautious in these high-risk patients.

All patients should be reviewed in the out-patient clinic 6 weeks after discharge if they have made a good recovery. They should generally stay within their home and avoid strenuous activity for the first week after discharge, only climbing stairs once a day if possible. Following this week, they should gradually increase their level of activity by taking daily walks of increasing length, so that by the time of their out-patient clinic visit they have returned to their normal pre-infarct level of activity. Sexual activity can be resumed after 3–4 weeks but may need to be modified to avoid strenuous activity initially. Leaflets relating to all aspects of rehabilitation are available from the British Heart Foundation and should be supplied to all patients.

Patients should be advised not to drive until the time of their first follow-up visit. It is the responsibility of the patient to ensure that their motor vehicle insurance is not compromised by their infarct. Private licences are retained, but PSV and HGV licences are automatically lost. In certain circumstances a PSV or HGV licence can be regained after an infarction. Details of the stringent criteria for return of these licences are available in a booklet obtainable from the DVLC at Swansea. If the patient has made a satisfactory recovery, they can resume driving following their out-patient attendance. Most patients should undergo an exercise test at the time of their first follow-up clinic visit to aid the selection of patients with a poor prognosis who require further investigation.

Patients who are in a good prognostic group should be discharged at the first 6-week follow-up visit. The presence of angina, heart failure, or adverse prognostic features are an indication for more intensive and prolonged investigation and follow-up. These patients may need to limit a previously active lifestyle or consider a change of occupation.

In patients who have had a more complicated course, mobilization may need to be slower, and out-patient follow-up earlier than 6 weeks.

Prognosis and risk stratification

In-hospital mortality can now be significantly reduced in patients with acute myocardial infarction by intensive early therapy with **aspirin**, **nitrates**, **beta-blockers**, and **thrombolytic therapy**. There remains a group of survivors who are at high risk of early death prior to discharge or in the next year.

In addition to clinial features known to be associated with a poor prognosis, a number of investigations can aid in risk stratification and subsequent management. Patients with evidence of poor ventricular function should have an echocardiogram, since anticoagulation is indicated if mural thrombus is present. A radionuclide ventriculogram, if available, should be carried out around 6 weeks post-infarction. Prognosis is dependent on residual left ventricular function, which can be accurately quantified by the radionuclide ejection fraction. Clinical, radionuclide, or echocardiographic evidence of poor left ventricular function in an infarct survivor should lead to institution of ACE inhibitor therapy since the SAVE study indicates that this will prevent ventricular dilatation and improve prognosis.

Patients at high risk of early death can be identified by using the following criteria;

- age greater than 60 years
- past history of ischaemic heart disease, hypertension, or diabetes
- evidence of extensive infarction on clinical, electrocardiographic, or enzyme criteria
- evidence of significant impairment of left ventricular function—i.e. overt heart failure, radionuclide ejection fraction below 30 per cent, cardiomegaly on chest X-ray
- persistent S–T-segment abnormalities, conduction defects, and sustained ventricular arrhythmias after the first 48 hours
- early post-infarct angina or infarct extension
- a post-infarct exercise test positive for ischaemia at low workload

Prognosis in these high-risk patients may be improved by medical intervention, and they should be considered for further investigation. Patients who sustain an uncomplicated small infarct, with no subsequent heart failure or angina, and who have a negative post-infarction exercise test, have a very good prognosis and require no further investigation.

Exercise testing

The question of when to exercise test post-infarct patients has not been completely resolved. Some physicians advocate the use of a submaximal test between days seven and ten. Although this provides important psychological support for the patient and their family, its value in detecting patients at high risk of death in the next 12 months is limited, since the test is not rigorous enough to reliably detect exercise-induced ventricular arrhythmias or ischaemia.

We recommend a full symptom-limited Bruce protocol exercise test at the time of the first clinic visit 6 weeks post-infarction. Patients who achieve a good level of exercise with a satisfactory haemodynamic response and no provokable arrhythmias or ischaemia are in a good prognostic group and require no further investigation.

Patients who have 2 mm or more of S–T depression or induced ventricular arrhythmias at a low workload (less than 6 minutes of the Bruce protocol) at the time of the 6-week exercise test should be considered for further investigation as they have a poor prognosis. Patients should only undergo exercise testing and angiography if they are fit enough for subsequent revascularization procedures.

Long-term beta-blockade

Long-term beta-blockade post-infarction has been shown to reduce subsequent mortality by 20–30 per cent. Provided there are no contraindications (asthma or chronic obstructive pulmonary disease, heart failure, AV block, hypotension, or resting bradycardia of less than 50 b.p.m.) to the use of beta-blockers, therapy should be started prior to discharge. Optimal duration of therapy is unknown, but treatment should be continued for at least 1 year. Various beta-blockers have been shown to be effective. The agent with the largest effect on mortality reduction in current studies is **timolol** in a dosage of 10 mg b.d.

Calcium antagonists

A recent overview of all the trials of calcium antagonists in myocardial infarction and unstable angina (involving more than 19 000 patients) concluded that there was no beneficial effect on infarct development, infarct size, reinfarction, or mortality. **Diltiazem** has been reported to be beneficial in the treatment of non-Q-wave infarction, but data is limited and inconclusive. Diltiazem may be beneficial in patients with well preserved ventricular function, but it is clearly harmful in those with significantly impaired ventricular function. **Verapamil** has recently been reported to have a beneficial effect on prognosis following myocardial infarction, but this positive result must be considered in the light of all the available negative results. At present, calcium antagonists cannot be recommended as routine prophylactic therapy in patients during or after acute myocardial infarction. **Verapamil** is probably the best available agent for secondary prophylaxis in patients in whom beta-blockers are contraindicated or poorly tolerated.

Hyperlipidaemia

Patients who survive for 5 years or more following an acute myocardial infarction show a reduction in late mortality with treatment of hyperlipidaemia. Patients who have a poor prognosis are not likely to survive for long enough to benefit from therapy. Lipid levels measured in the first 24 hours post-infarction are thought to accurately reflect pre-infarct levels and can therefore be used to assess the need for subsequent lipid-lowering therapy. After the first 24 hours, lipid levels are depressed; if a sample is not available until later, decisions on the need for lipid-lowering therapy should be deferred until an estimate can be obtained at 3 months, when levels will have returned to the pre-infarct state.

Cholesterol levels are not greatly affected by the timing of meals. A random sample taken within 24 hours of the onset of infarction is, therefore, adequate for initial assessment. Triglyceride levels are influenced by the timing of meals, and if elevated on an admission sample the measurement should be repeated in the fasting state 3 months post-infarction before decisions on therapy are made. Dietary therapy can produce a 5–15 per cent reduction in serum cholesterol. First-line drug therapy remains a resin (such as **cholestyramine**) for patients with isolated elevation of cholesterol or a fibrate (such as **bezafibrate**) for patients with a mixed elevation of both cholesterol and triglycerides. HMG CoA reductase inhibitors are more effective than the older drugs, producing a 20–40 per cent fall in lipid levels. Although these drugs may well replace resins and fibrates as first-line therapy in the near future, they are still undergoing clinical assessment, and at present we still recommend initial treatment of hyperlipidaemia with the older agents. Our current policy is:

- Do not treat hyperlipidaemia aggressively in post-infarct patients who have a poor prognosis, as they are unlikely to survive long enough to benefit from therapy. Dietary modification is preferable to drug therapy if treatment is considered appropriate.

- For patients with a good prognosis and a cholesterol level of 6.5–7.8 mmol/l, institute dietary therapy and review with a repeat lipid level after 6 months. If cholesterol remains above 6.5 mmol/l, consider therapy with a resin or a fibrate initially. If these agents fail, consider an HMG CoA reductase inhibitor.
- For patients with a good prognosis who have an initial cholesterol above 7.8 mmol/l dietary therapy is unlikely to reduce their cholesterol below the target level of 6.5 mmol/l. Initial therapy should be with both diet and a resin or a fibrate. If these measures fail, consider an HMG CoA reductase inhibitor.

In young patients, those with very high lipid levels, or those who fail to respond to the above measures, consider referral to a lipid clinic. Remember to screen first-degree relatives of patients with severe hyperlipidaemia, as this may be familial. Secondary causes of hyperlipidaemia (such as renal failure, thyroid dysfunction, drug therapy, or alcohol abuse) should be excluded before discharge.

Hypertension

Control of hypertension that persists following recovery from myocardial infarction is indicated. Treatment is likely to reduce subsequent mortality from stroke and heart failure, though evidence that it reduces death rates from ischaemic heart disease is scanty. **Beta-blockers** are the agents of first choice, since they are known to have beneficial effects on prognosis in the post-infarction population. Control of hypertension should not aim to reduce blood pressure below a level of approximately 130/80 mmHg, as there may be a J-shaped relationship between blood pressure and cardiovascular mortality.

Indications for revascularization

Patients whose unstable angina fails to settle with maximal medical therapy are likely to progress to infarction. These patients should undergo early angiography with a view to PTCA or bypass grafting, depending on their coronary anatomy.

Patients with evidence of continuing ischaemia following infarction (i.e. ischaemic pain at rest or on minimal exertion despite antianginal therapy) or an exercise test positive at low workload should undergo angiography with a view to revascularization for both symptomatic control and on prognostic grounds since patients with left main stem disease or three vessel disease and impaired left ventricular function have an improved survival after coronary artery bypass grafting.

Lifestyle modification

Patients who survive a myocardial infarct or an episode of unstable angina should be strongly advised to stop smoking, as recent studies have demonstrated that this has a beneficial effect on mortality in both men and women.

There is little hard data to substantiate improvement in long-term prognosis with structured exercise rehabilitation programmes after infarction, though they help to improve the sense of well-being, and assist the patients in modifying risk factors favourably. If a rehabilitation programme is available locally, referral of all infarct survivors other than those who have sustained a large anterior Q-wave infarction is desirable.

Key points

- Patients with clinical features indicative of a poor prognosis will require more intensive treatment and follow up.

- Medical therapy may improve prognosis in survivors of extensive infarction (i.e. ACE inhibitors and **anti-coagulants**).
- Evidence of persistent ischaemia, infarct extension, or a positive exercise test should lead to angiography with a view to revascularization.
- Long-term beta-blockers should be routinely used unless a contraindication is present. Prophylactic calcium antagonists are not routinely recommended.
- Treatment of hyperlipidaemia may improve long-term prognosis in post-infarct patients who survive for 5 years or longer.

Further reading

Bonow, R. O., Kent, K. M., Rosing, D. R., *et al.* (1984). Exercise induced ischaemia in mildly symptomatic patients with coronary artery disease and preserved left ventricular function: identification of subgroups at risk of death during medical therapy. *N. Engl. J. Med.*, **311**, 1339–45.

Campbell, S., A'Hern, R., Quigley, P., Vincent, R., Jewitt, D., and Chamberlain, D. (1988). Identification of patients at low risk of dying after acute myocardial infarction, by simple clinical and submaximal exercise test criteria. *Eur. Heart J.*, **9**, 938–47.

Cohn, J. N. (1992). The prevention of heart failure—a new agenda. *N. Engl. J. Med.*, **327**, 725–7.

Gold, R. and Oliver, M. (1989). Fitness to drive: update guidance on cardiac conditions in vocational and other professional drivers. *Health Trends*, **21**, 88–9.

Gold, R. and Oliver, M. (1990). Fitness to drive: updated guidance on cardiac conditions in holders of ordinary driving licences. *Health Trends*, **22**, 31–2.

Held, P. H., Yusef, S., and Furberg, C. D. (1989). Calcium channel blockers in acute myocardial infarction and unstable angina: an overview. *Br. Med. J.*, **299**, 1187–92.

Mark, D. B., Hlatky, M. A., Harrel Jr, F. E., Lee, K. L., Califf, R. M., Pryor, D. B. (1987). Exercise treadmill score for predicting prognosis in coronary artery disease. *Ann. Intern. Med.*, **106,** 793–800.

Moss, A. J. and Benhorin, J. (1990). Prognosis and management after a first myocardial infarction. *N. Engl. J. Med.*, **322,** 743–53.

Oldridge, N. B., Guyatt, G. H., Fischer, M. S., and Rimm, A. A. (1988). Cardiac rehabilitation after myocardial infarction. Combined experience of randomised clinical trials. *J.A.M.A.*, **260,** 945–50.

Rautaharju, P. M., Prineas, R. J., Eifler, W. J., *et al.* (1986). Prognostic value of exercise electrocardiogram in men at high risk of future coronary heart disease: multiple risk factor intervention trial experience. *J. Am. Coll. Cardiol.*, **8,** 1–10.

The Norwegian Multicenter Study Group (1981). Timolol induced reduction in mortality and reinfarction patients surviving acute myocardial infarction. *N. Engl. J. Med.*, **304,** 801–7.

Weiner, D. A., Ryan, T. J., McCabe, C. H., *et al.* (1987). Value of exercise testing in determining the risk classification and the response to coronary artery bypass grafting in three-vessel coronary artery disease: a report from the Coronary Artery Surgery Survey (CASS) Registry. *Am. J. Cardiol.*, **60,** 262–6.

Yusef, S., Peto, R., Lewis, J., Collins, R. and Sleight, P. (1985). Beta blockade during and after myocardial infarction: an overview of the randomised trials. *Prog. Cardiovasc. Dis.*, **27,** 335–71.

NOTES

NOTES

NOTES

NOTES

CHAPTER 15

Non-ischaemic cardiovascular emergencies

Aortic dissection

Aortic dissection occurs at a rate of approximately ten per million per year. It is twice as common in males and has a peak incidence in the sixth and seventh decades. Most dissections arise in aortae which have medial degeneration due to hypertension (the major risk factor in over 50 per cent of cases). Other well known associations are Marfan's syndrome, vasculitic disorders affecting the aorta, pregnancy, coarctation, and bicuspid aortic valve. The majority (70 per cent) of acute dissections arise in the ascending aorta (the area subject to maximum haemodynamic stress) with the remaining 30 per cent arising distal to the left subclavian artery.

The initial event is the sudden development of an intimal tear in the aorta. As the tear develops, blood at systemic arterial pressure tracks into the underlying degenerative media producing the acute dissection, which then propagates for varying distances throughout the aorta. Rupture of the haematoma into the pleural or pericardial cavity leads to rapid death in the majority of patients before they reach hospital.

The diagnosis of acute dissection should be considered when:

- Risk factors such as hypertension or a connective tissue disorder are present.
- The presenting chest pain is very severe, of sudden onset, tearing in nature, or radiates to the back.
- Pulse deficits, neurological abnormalities, a difference in blood pressure in the arms, or aortic regurgitation are present.
- The admission chest X-ray is abnormal showing a wide upper mediastinum.
- The admission ECG does not show evidence of acute infarction in a patient with severe chest pain. Remember that a proximal dissection can involve the coronary ostia and produce an associated infarct. Do not, therefore, rule out the diagnosis of aortic dissection simply because the ECG shows signs of an infarct.

If you suspect the presence of an aortic dissection the diagnosis can be confirmed by:

- Echocardiography. Traditional transthoracic echocardiography is unreliable and does not exclude the diagnosis; transoesophageal echocardiography is sensitive and specific but not widely available as yet.
- CT scanning.
- Aortography.

When the diagnosis is suspected in a unit where access to appropriate diagnostic techniques are unavailable, emergency treatment is directed towards reducing blood pressure and myocardial contractility, controlling pain, and transferring the patient to a unit where a definitive diagnosis and appropriate surgical treatment, if indicated, can be arranged. Initial therapy is:

1. IV **diamorphine** 2.5 mg at 1 mg/min initially plus **metoclopramide** to relieve pain.
2. IV beta-blockers to reduce myocardial contractility, i.e. IV **propranolol** 0.1 mg/kg every 4–6 minutes until heart rate drops to below 60 b.p.m.
3. Reduce systolic BP to 100–120 mmHg with **sodium nitroprusside** initially at 10 μg/min, increasing in steps of 10 μg/min every 5 minutes until blood pressure drops to the required range (see Chapter 18 for nitroprusside regime).

Sodium nitroprusside should ideally be used in conjunction with intra-arterial blood-pressure monitoring. If this is not available, alternative effective regimes are:

- Intravenous **labetalol** instead of **propanolol** and **nitroprusside**. The regime is identical to that used in hypertensive emergencies and is detailed in Chapter 18.
- A combination of a beta-blocker and an intravenous **nitrate**. The beta-blocker should be given first to prevent drug-induced tachycardia. A **GTN** or **isosorbide** infusion should then be commenced at a low initial dose, increasing every 15 minutes until the blood pressure falls to the desired level.

Once a definitive diagnosis is made, surgery is recommended for proximal dissections, and continued medical therapy for distal dissections.

Remember:

If you have any suspicion that an acute dissection is present, do not give thrombolytic therapy: institute first-line therapy with diamorphine, propranolol, and nitroprusside and arrange investigations to confirm or refute the diagnosis as a matter of urgency.

Severe hypertension

Severe uncontrolled hypertension is not commonly seen in the 1990s. The syndrome is characterized by:

- diastolic pressures above 120 mmHg
- severe (grade III or IV) retinopathy
- proteinuria

The aim of treatment should be to reduce blood pressure to a level of around 180/110 over several hours, since abrupt reduction of blood pressure can precipitate stroke, myocardial infarction, or renal failure. Oral therapy is preferable, and the only indications for IV treatment are:

- aortic dissection
- acute severe hypertension in phaeochromocytoma or monoamine oxidase inhibitor overdose
- hypertensive left ventricular failure
- hypertensive encephalopathy

The IV drugs of choice for hypertensive emergencies are **labetalol** (see Chapter 18 for regime) or **nitroprusside** if a beta-blocker is contraindicated. Concomitant oral therapy should be instituted along with intravenous therapy. Intravenous **frusemide** may be necessary if left ventricular failure is present. As previously noted, the combination of a beta-blocker and a nitrate infusion may be preferable to the use of nitroprusside. The management of a hypertensive crisis is a difficult area, and a senior colleague should be consulted before undertaking treatment.

Cardiac tamponade

Cardiac tamponade occurs when blood or other fluid fills the pericardial space, raising intrapericardial pressure, compressing the heart and preventing it from filling. As the heart is unable to fill adequately, stroke volume, blood pressure, and cardiac output decrease and signs of shock with a raised venous pressure develop.

There are many possible causes of cardiac tamponade. The commonest seen on the CCU is rupture of the free wall of the ventricle following acute infarction, which is usually rapidly fatal. Occasionally a patient will be seen who develops tamponade following cardiac catheterization or temporary pacing. Malignant pericardial effusions causing tamponade are more common, but tend to have a less acute presentation. You should have a high index of suspicion that tamponade may be present in a haemodynamically collapsed patient with renal failure, post-cardiac surgery, and following chest injuries (particularly in patients treated with anticoagulants).

The major clinical signs of tamponade are:

- hypotension with pulsus paradoxus
- raised venous pressure with positive Kussmaul's sign
- soft heart sounds

ECG and chest X-ray signs (a globular heart, electrical alternans, and low-voltage QRS complexes) may be present. The best way to make a definitive diagnosis is echocardiography. Treatment of a large effusion causing tamponade is by pericardiocentesis, described in Chapter 17.

Digoxin toxicity

Toxicity should be suspected in any patient on **digoxin** who develops:

- anorexia
- nausea
- vomiting

- diarrhoea
- confusion
- visual disturbance

Toxicity is most likely to occur when hypokalaemia is also present, in the elderly, or in patients with impaired renal function. A serum digoxin level of greater than 3.0 ng/ml (3.8 nmol/l) supports the clinical diagnosis.

If you suspect digoxin toxicity:

1. Stop digoxin.
2. Correct hypokalaemia if present.
3. Correct any arrhythmia that occurs.

Atrial tachycardia with AV block is the commonest arrhythmia. Treatment is with **beta-blockers** if the ventricular rate is fast (as detailed in Chapter 7). If VT develops, **phenytoin** is the drug of choice (given as a bolus of 100 mg over 2 minutes, which can be repeated every 5 minutes up to a maximum of 1000 mg if unacceptable side effects do not occur). **Lignocaine** and **phenytoin** should not be used concurrently.

Remember to avoid DC cardioversion for the treatment of tachyarrhythmias in a patient with digoxin toxicity, as it may induce intractable arrhythmias. If you have to give a DC shock in these circumstances, start at a low energy level (i.e. 10 J). If **digoxin** toxicity occurs due to an overdose of **digoxin**, treatment with **digoxin antibody fragments** in addition to the above measures may be life-saving.

Key points

- Do not attempts to rapidly correct hypertension over a short time period in patients with hypertensive emergencies.

- Consider pericardial effusion in a haemodynamically collapsed patient who has risk factors.
- Consider the diagnosis of aortic dissection in a patient with an atypical history of chest pain, particularly if risk factors are present.
- Aortic dissection is best managed in a unit with access to diagnostic facilities and cardiac surgery. If these are not available to you, institute appropriate medical therapy and arrange transfer to the local invasive unit.

Further reading

De Sanctis, R. W., Doroghazi, R. M., Austen, W. G., and Buckley, M. J. (1987). Aortic dissection. *N. Engl. J. Med.*, **317,** 1060–7.

Ferguson, R. K. and Vlasses, P. H. (1986). Hypertensive emergencies and urgencies. *J.A.M.A.*, **255,** 1607–13.

Horgan, J. H. (1987). Cardiac tamponade. *Br. Med. J.*, **295,** 563–4.

Houston, M. C. (1987). Pathophysiology, clinical aspects, and treatment of hypertensive crises. *Prog. Cardiovasc. Dis.*, **32,** 99–148.

Lancet (1988). Acute aortic dissection. *Lancet*, **ii,** 827–8.

Markiewicz, W., Borovik, R., and Ecker, S. (1986). Cardiac tamponade in medical patients: treatment and prognosis in the echocardiographic era. *Am. Heart. J.*, **111,** 1138–42.

Smith, T. W., Haber, E., Yeatman, L., and Butler, V. P. (1976). Reversal of advanced digoxin intoxication with Fab fragments of digoxin specific antibodies. *N. Engl. J. Med.*, **294,** 797–800.

NOTES

NOTES

NOTES

CHAPTER 16

Commonly used cardiac drugs

Beta-blockers

- These remain first-line therapy for the treatment of hypertension and angina, and have a role in the treatment of acute infarction, secondary prophylaxis post-infarction, and arrhythmia control.
- Many different drugs are available. We recommend use of a relatively short-acting cardioselective drug such as **metoprolol** 50 mg twice daily.
- A wide range of side-effects can occur, but the drugs are generally well tolerated.

Calcium antagonists

- These can now be considered to be acceptable therapy for hypertension and are useful in angina, where a beta-blocker alone is ineffective or not tolerated.

- They have an advantage over beta-blockers in that they do not cause lipid elevation.
- Up to 20 per cent of patients may develop side-effects (oedema, flushing, headache) necessitating withdrawal of therapy.
- There is little to choose between the various calcium antagonists. It is our policy to use **nifedipine** 10–20 mg t.d.s. for angina, as it can safely be used in conjunction with a beta blocker and **nifedipine SR** in hypertension (where its twice-daily administration is an advantage).
- **Diltiazem** is a useful alternative calcium antagonist with a low incidence of side-effects. It is similar to **verapamil** in its effect on nodal conduction, and is a useful first line agent for the treatment of angina in patients in whom beta-blockers are contraindicated or poorly tolerated. **Verapamil** and **Diltiazem** should not be used with a beta-blocker, since this combination may cause profound bradycardia.

Nitrates

- **GTN** 600 μg sublingually remains standard therapy for the treatment of anginal attacks. Advising patients to use their tablets prophylactically will often obviate the need for multi-drug regimes with their attendant high incidence of side-effects.
- Aerosol nitrate preparations tend to work quicker than GTN tablets, but their major advantage is that the GTN is not degraded (the shelf life of GTN tablets is only 6 weeks) and aerosols are therefore preferable for patients who use nitrates infrequently.
- Long-acting nitrates have been a subject of great controversy. It is now clear that nitrate tolerance is a common problem, and that early regimes were faulty.

We recommend:

1. Oral nitrate preparations to be given twice daily, in the morning and early afternoon (i.e. **isosorbide mononitrate** 10 mg b.d., at 0800 and 1300) to allow levels to drop later in the day and provide a nitrate-free interval.
2. Remove GTN patches for 8 hours of the day to allow a nitrate-free interval.
3. IV nitrates are useful for in-hospital therapy. Infusion rates can be carefully titrated to achieve optimal clinical responses. Tolerance is a problem, and interrupted regimes or an increase in dosage may be necessary if the infusion continues for more than 24 hours.

Aspirin

- The use of **aspirin** as secondary prophylaxis after a TIA, myocardial infarction, episode of unstable angina, or stroke is of proven benefit in reducing the incidence of subsequent vascular events.
- Aspirin as primary prophylaxis in patients at high risk of vascular events is not of proven benefit, and we do not advocate its use in this situation.
- The best dose to use is controversial. In theory, lower doses are most beneficial (maximum inhibition of platelet aggregation with minimal reduction in endothelial prostacyclin production) and have the lowest incidence of GI side-effects, but data on their use is limited. The doses of proven efficiency are 160 mg daily in acute MI and 325 mg daily in unstable angina. A recent study of unstable angina and non-Q-wave infarction suggested that 75 mg daily may also be effective in such patients, and this regime may therefore be a viable alternative for those patients who are intolerant of higher doses.

Inotropic agents

- It is our view that **digoxin** is now rarely indicated in the treatment of heart failure with sinus rhythm. It is recommended only where therapy with diuretics, and ACE inhibitors has failed, and the patient has a dilated, poorly functioning LV. It is still indicated for the treatment of supraventricular arrhythmias.
- Toxicity is common with digoxin and the therapeutic window is narrow. Reduction in dosage is necessary in the elderly or those with impaired renal function. Remember interactions with other cardiac drugs (**amiodarone**, **quinidine**, and **verapamil** increase plasma digoxin levels) and the danger of diuretic-induced hypokalaemia.
- **Dobutamine** in the dosage range 2.5–20 μg/kg/min is our IV inotropic agent of choice, used to support the circulation in refractory heart failure.
- **Dopamine** is most useful as a renal vasodilator, used to preserve renal function in low output states in the dose range 2.5–5.0 μg/kg/min. At doses above this level it provides inotropic support, but its propensity to cause arrhythmias is high. **Dopamine** and **dobutamine** are useful if used together to support the circulation and preserve renal function. Remember that **dobutamine** can be given via a peripheral line, but **dopamine** must be given via a central line.
- New phosphodiesterase inhibitors (i.e. **milrinone** or **enoximone**) have both inotropic and vasodilating properties. They may prove to have a role in acute heart failure, but data as yet is limited. Their main drawback is their tendency to cause hypotension, which limits their use in cardiogenic shock.
- **Noradrenaline** is a powerful vasoconstrictor and can be used in patients who remain hypotensive despite

maximal supportive therapy with other agents. Its use is often associated with tachycardia and ventricular arrhythmias, and it is not therefore first-line therapy.

Diuretics

- Thiazide diuretics can be used for mild heart failure. They are useful in the elderly as their less dramatic diuretic effect is easier to cope with for frail and relatively immobile patients who often have pre-existing urinary-tract problems.
- In the treatment of hypertension, thiazide diuretics are of proven efficacy and are inexpensive. They do have an adverse effect on potassium and lipid metabolism, and this may account for their inability to reduce ischaemic heart disease mortality in hypertensives. Beta-blockers do have some effect on ischaemic heart disease mortality in some hypertensive subgroups, and are therefore probably first-choice antihypertensive agents. Calcium antagonists and ACE inhibitors are well tolerated in hypertension, but their effect on outcome has not been extensively studied, and they are at present second-line agents.
- Proprietary preparations containing small amounts of potassium are not recommended, as the amount of potassium in the combination is not adequate. A combination of a thiazide with a potassium-retaining agent (i.e. **dyazide**) is preferable.
- For any but the mildest heart failure, a loop diuretic such as **frusemide** in doses of 40 mg daily or more will be required. Again, this should always be prescribed with a potassium-retaining agent (i.e. **Frumil** containing 40 mg of frusemide and 5 mg of **amiloride** per tablet), as diuretic-induced hypokalaemia may

potentiate rhythm disturbances in these patients, many of whom die suddenly from ventricular arrhythmias.

- Remember that when a patient's diuretic requirement increases above 40 mg of frusemide daily, it is time to consider introducing ACE inhibitors.
- Patients who develop severe congestive heart failure will stop absorbing their oral drugs due to mucosal oedema in the GI tract. Changing these patients to IV diuretics often initiates a brisk diuresis which can be maintained with oral agents once the oedema has cleared and GI absorption is improved.
- Heart failure refractory to maximal conventional therapy may respond to the addition of a thiazide diuretic. We recommend **metolazone** 2.5–5.0 mg daily; careful monitoring of daily weight and electrolytes is necessary in the face of the very brisk diuresis that this regime usually produces. Great caution should be exercised in sending patients home on this agent, as they are at risk of severe dehydration and pre-renal uraemia.

ACE inhibitors

- These drugs have revolutionized the treatment of heart failure. Their propensity to cause side-effects has been exaggerated, and if used in properly selected patients and appropriate doses, serious side-effects are rare.
- Their relative freedom from unpleasant side-effects may lead to their future use as first-line agents in the treatment of hypertension. At present we prefer to use beta-blockers, since these are of proven efficacy in reducing stroke mortality.

- **Captopril** is preferable for initiating therapy, as the duration of any hypotensive response that occurs with the first dose is shorter than with other long-acting drugs. **Enalapril** and **lisinopril** have the advantage of once-daily administration, and it is reasonable to switch patients to comparable doses of a long-acting drug once they are established on captopril to aid compliance.
- ACE inhibitors have been shown to reduce mortality in mild and severe heart failure. No other agents (other than the combination of hydralazine and isosorbide in a single trial with a high drop-out rate due to side-effects) have been shown to do this. Caution should be exercised when introducing ACE inhibitors in a patient when:
  1. There is evidence of widespread vascular disease. These patients have a high incidence of renal artery stenosis, and ACE inhibitors may precipitate occlusion of the stenosed vessel.
  2. Large doses of diuretics are being administered and the patient is relatively dehydrated or hyponatraemic. Diuretics should be omitted for 36 hours before the induction of therapy in these patients.
- Current trends in heart failure therapy are to introduce ACE inhibitors at an early stage (i.e. once diuretic requirement rises above 40 mg of **frusemide**) as there is now clear evidence from the SAVE and SOLVD studies that they may slow progression of ventricular dilatation.
- ACE inhibitors can have an adverse effect on renal function. We advise checking baseline renal function prior to starting therapy, and repeating checks at regular intervals after initiation of therapy to detect drug-induced deterioration in renal function.

Antiarrhythmic agents

- There are a multitude of antiarrhythmic agents available. It is better to have a sound knowledge of the use of four or five agents and to stick to these.
- The use of **digoxin** for control of supraventricular arrhythmias is described elsewhere. It remains the drug of first choice for control of atrial fibrillation or flutter. If digoxin alone fails, then addition of **atenolol**, **verapamil** (in those with good LV function), or **amiodarone** (in those with impaired LV function) should be considered.
- **Adenosine** will terminate most SVTs and aid the diagnosis of broad complex tachycardias.
- For ventricular arrhythmias **lignocaine** is the drug of first choice. If long-term oral therapy is required, another Class I agent such as **mexilitine** or **disopyramide** can be used. Remember that all Class I agents are negatively inotropic.
- For refractory supraventricular or ventricular arrhythmias when other agents have failed **amiodarone** is often effective. The problem is its wide range of serious side-effects. Patients on long-term oral therapy should have TFTs, LFTs, and lung function monitored carefully. One of **amiodarone**'s major advantages is its relative lack of negative inotropic effects. It is therefore a useful drug when Class I, II, or IV agents are contraindicated.
- **Verapamil** is useful for the control of paroxysmal SVT in patients with normal ventricular function in whom **adenosine** is contraindicated, but its use IV in any other situation is strongly discouraged.

Key points

- Try to become familiar with the use of one agent in each class of drugs.
- Use asymmetrical dosing regimes for **nitrates** to prevent tolerance.
- **Digoxin** is probably not of great benefit to the patient with heart failure in sinus rhythm, and should only be used when other therapies have failed to control symptoms.
- ACE inhibitors are known to improve signs, symptoms, and prognosis in patients with impaired ventricular function, and should be used from an early stage.
- Digoxin, **beta-blockers**, **adenosine**, and **verapamil** will control most SVTs. **Lignocaine** and **amiodarone** will control most ventricular arrhythmias.

Further reading

Hammer, J. (1987). *Drugs for heart disease*. Chapman and Hall, London.

Opie, L. H. (1991). Drugs for the heart. W. B. Saunders, Philadelphia.

NOTES

NOTES

NOTES

CHAPTER 17

Practical procedures

No written guidelines can replace instruction by an experienced colleague. These short notes are meant to be used as an adjunct to this and not as a replacement.

Central venous cannulation

Cannulation of a central vein is used on the CCU to:

- Monitor central cardiac pressures to guide therapy.
- Insert temporary pacing wires.
- Provide a route for administration of drugs such as **dopamine** that should not be given via a peripheral line.
- Provide intravenous access in a patient with no suitable peripheral veins.

A number of sites can be used to provide central venous access. In general, we use the right subclavian unless the patient has a bleeding tendency, when we use the internal jugular, antecubital, or femoral venous route, since these allow easier haemostasis to be achieved.

For subclavian cannulation using a Seldinger technique:

1. Lie the patient flat, with the head tilted down if possible
2. Clean and drape the skin.
3. Infiltrate local anaesthetic below the junction of the medial one-third and lateral two-thirds of the clavicle and nick the skin at this point with a scalpel blade.
4. Straighten the clavicle by having an assistant apply gentle traction to the patient's right arm.
5. Place your left index finger in the suprasternal notch and advance the introducer needle from the skin nick towards your finger until you come into contact with the clavicle. Move the tip of the needle down until it lies just below the clavicle, then slowly advance the needle towards your finger in the suprasternal notch while aspirating for blood. Avoid pointing the tip of the needle downwards to minimize the risk of an iatrogenic pneumothorax.
6. When the vein is punctured, remove the syringe (covering the needle with your thumb to prevent entry of air) and insert the guide-wire, flexible end first. The wire should enter freely with no resistance; if you encounter resistance remove the wire and check that the tip of the needle is still in the vein by aspirating more blood, then try again.
7. When the guide-wire is in place, remove the needle and insert the dilator and cannula over the wire, then remove the guide-wire and dilator, leaving the cannula in place.
8. Proceed to place the pacing wire, pulmonary artery

catheter, or attach intravenous infusion, then secure cannula.

9. Check a post-procedure chest X-ray to exclude pneumothorax.

Temporary pacing

1. Before starting, ensure that the resuscitation trolley is easily accessible and that you have venous access to give drugs if necessary. Brady- or tachyarrhythmia are often induced during the procedure so ensure that **atropine**, **isoprenaline**, and **lignocaine** are ready to hand in case they are needed.
2. The best route for placing a temporary wire is the subclavian vein. If subclavian puncture is contra-indicated, use the femoral or internal jugular; these have the advantage that haemostasis can easily be secured by pressure if bleeding occurs.
3. A sheath with a non-return valve should be placed in the vein using a Seldinger technique to allow easy manipulation of the wire.
4. A size 5 or 6 bipolar pacing wire is appropriate in most cases. Ensure that a working pacing box is ready before inserting the wire.
5. Arrhythmias are often induced as the wire crosses the tricuspid valve. These usually settle rapidly and require no treatment.
6. The best position for the pacing wire is usually with the tip of the wire in the right ventricular apex, to the left of the midline, with the tip of the wire pointing inferiorly on screening.
7. Acceptable stimulation thresholds are below 1.0 V, although higher levels may be acceptable if the patient is elderly, has had an inferior infarct, or

several sites have been tried all with relatively high thresholds.

8. Check stability with the pacing wire switched on at an output voltage twice the threshold and the patient deep breathing. If capture is lost during this manoeuvre then try to find a more stable position.

9. Remember to secure the wire firmly with a loop formed on the chest wall to minimize the chance of its being pulled out easily.

10. Pacing threshold should be checked twice daily. A sudden increase in threshold usually indicates the need for repositioning. Pacing thresholds are checked by increasing the pacing rate to obtain continuous pacing, then progressively decreasing the output voltage until capture is lost (Fig. 17.1). The minimum output voltage which provides consistent capture is the pacing threshold voltage. Once the threshold has been ascertained, the output voltage should be set at three times the threshold to ensure a margin of safety.

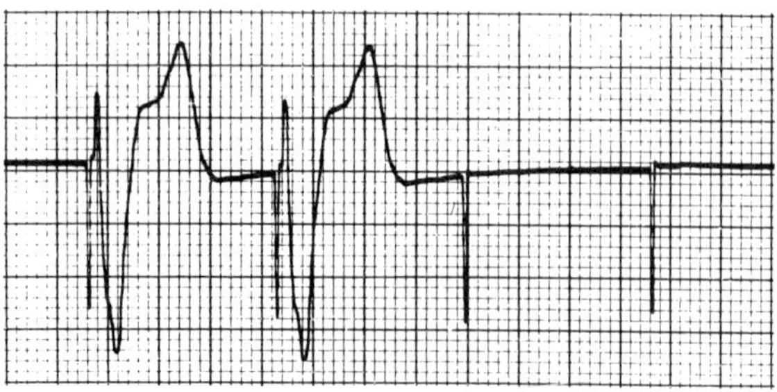

Fig. 17.1 Determination of pacing threshold. First two complexes show normal capture at 0.5 V. Output voltage is then reduced to 0.4 V and the subsequent pacing spikes fail to capture the ventricle

Pulmonary artery catheterization

Pulmonary artery balloon flotation catheters enable central cardiac pressures and cardiac output to be monitored. This information can be very useful in diagnosis and treatment, but insertion of these lines is not without risk of complications such as pneumothorax, infection, or arrhythmia. The procedure should, therefore, only be carried out where the information provided will be of definite benefit to the patient.

Indications for use

- Low output states post-MI. These may be due to cardiogenic shock, hypovolaemia, or right ventricular infarction. Measurement of central pressures enables differentiation and appropriate treatment.
- Differentiation of cardiogenic pulmonary oedema (raised wedge pressure) from other causes of diffuse alveolar shadowing (i.e. in adult respiratory distress syndrome the wedge pressure is low or normal).

Placement of the catheter

1. Insert a suitable central venous sheath.
2. Check that the balloon inflates before insertion.
3. Flush the catheter with saline, connect to the monitoring system, and check that pressure is being recorded by moving the tip of the catheter rapidly up and down and observing simultaneous deflection of the pressure trace.
4. Insert the catheter 10 cm (circumferential lines on the catheter are 10 cm apart) and inflate the balloon. Advance the catheter using X-ray guidance to the pulmonary artery. If X-ray guidance is not available, observation of the pressure trace can be used to guide catheter placement. Do not attempt to place a

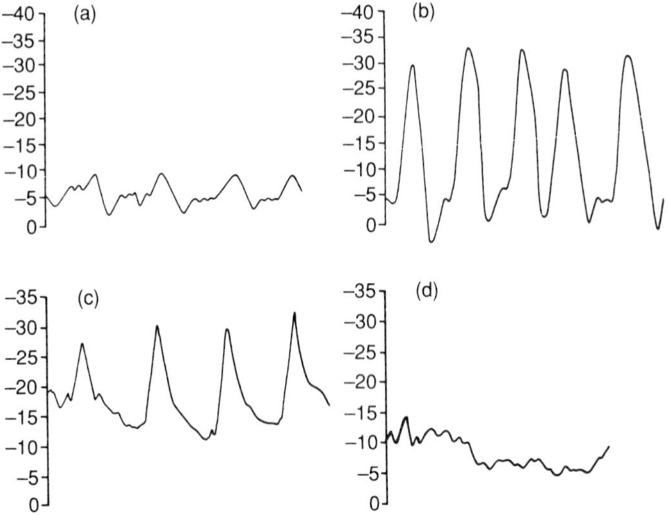

Fig. 17.2 Normal right-sided intracardiac pressure tracings obtained with a pulmonary artery balloon flotation catheter. (a) is the right atrium, (b) the right ventricle, (c) the main pulmonary artery, and (d) the pulmonary wedge trace

catheter without X-ray guidance unless you have done so before under supervision. Figure 17.2 shows typical pressure tracings obtained from the different cardiac chambers.

5. Ventricular arrhythmias (usually runs of ectopics) are common as the catheter crosses the tricuspid valve.
6. When a wedge trace is acquired, deflate the balloon. A phasic pulmonary artery trace should reappear; if it does not, the catheter should be withdrawn slightly and the balloon reinflated and deflated until it is possible to obtain both a wedge and phasic pulmonary artery trace with the balloon fully inflated and deflated. The catheter is now in an optimal position.

Potential problems

- Pressure trace damped. This is usually due to kinking of the catheter, air bubbles in the pressure monitor, or thrombus occluding the catheter tip and can usually be solved by checking the pressure monitoring system for bubbles or kinks, or flushing the catheter.
- Failure to wedge. This is due to the catheter being placed too proximally or balloon rupture. If the balloon ruptures, blood will collect in the inflation syringe. The solution is to reposition a working catheter (advancing it until balloon inflation produces a wedge trace) or replace a catheter with a ruptured balloon.
- Whenever possible do not leave a catheter *in situ* for more than 48 hours to minimize the risk of infection. If infection occurs, remove the catheter, take blood cultures, and treat initially with broad spectrum IV antibiotics.
- Do not leave the balloon inflated for long periods, as this can cause pulmonary infarction.

Normal range of intracardiac pressure (measured from mid-chest)

Site	Pressure (mmHg)
RA	0–7
RV	15–25/0–8
PA	15–25/8–15
PAWP	6–12

Note—Wedge pressure is normally measured at the end of expiration. In a patient with COPD, pressures may be falsely elevated. In a patient without COPD, pulmonary artery diastolic pressure is normally equivalent to the wed pressure.

Pericardial aspiration

Pericardial aspiration is a potentially dangerous procedure; only attempt it if a senior colleague is available to supervise you. Guidelines for pericardial aspiration are set out below:

1. Pericardial aspiration should ideally be performed with full X-ray screening, rhythm monitoring, and resuscitation facilities readily available. If available, echocardiography can be used to guide pericardial aspiration since it allows you to visualize the fluid-filled space, the myocardium, and the aspiration needle.
2. Place the patient at a 30–45° angle.
3. Connect the V_1 lead from the ECG machine via a clip to the aspiration needle. If you touch the myocardium with the needle, S–T elevation will immediately show on the monitor giving you warning to stop advancing the needle.
4. Infiltrate **lignocaine** below and to the left of the xiphoid process.
5. Advance the needle towards the left shoulder, aspirating as you do so. Stop if S–T elevation develops on the monitor, as soon as you aspirate pericardial fluid, or when the needle is seen to enter the pericardial space on the echocardiogram.
6. Insert a guide-wire under radiological control then a pigtail drainage catheter. The catheter should be left *in situ* if there is any possibility of the fluid reaccumulating.
7. If echocardiographic screening is available, the apical route may be preferable since this reduces the chance of traumatizing the right ventricle. You should only attempt this if you have previous experience or the assistance of a senior colleague.

DC cardioversion

DC cardioversion is used to treat supraventricular and ventricular arrhythmias which are unresponsive to drug therapy or associated with haemodynamic collapse.

Remember:

DC cardioversion should be avoided if digoxin toxicity is present, as it may induce refractory ventricular arrhythmias or asystole.

Lignocaine, **phenytoin**, **beta-blockers**, and overdrive pacing can all be used to terminate **digoxin**-induced arrhythmias. DC cardioversion is safe in a patient on therapeutic doses of **digoxin** who is not toxic. **Lignocaine** and **phenytoin** should not be used in the same patient.

In a patient with VF, consciousness is usually rapidly lost. A non-synchronized shock is given, initially of 200 J, increasing to 360 J in unresponsive cases. For cardioversion of other acute arrhythmias, consciousness may not be lost, and the patient should be sedated with 5–10 mg of intravenous **diazepam** or a short-acting general anaesthetic. A synchronized DC shock is then given. Low energies may suffice in supraventricular arrhythmias, particularly atrial flutter, when 25 J may be successful. For elective cardioversions of out-patients, full oral anticoagulation with **warfarin** for at least 1 month prior to and following the procedure is necessary.

Echocardiography

Cardiac ultrasound machines have facilities for two dimensional, M-mode and Doppler echocardiography. Echocardiography is particularly useful, when available, in the following situations:

- Low output states or hypotension to assess left ventricular function, and to exclude mitral regurgitation, ventricular septal rupture, or pericardial effusion.

- Suspected aortic dissection. Conventional transthoracic echocardiography may demonstrate an abnormal dilated aortic root or aortic regurgitation. Trans-oesophageal echocardiography, if available, may be the first-choice diagnostic technique.

Key points

- Obtain instruction and supervision from an experienced colleague in all practical procedures.
- If placement of a central venous line proves difficult, do not persist in your attempts—ask a senior colleague for help.
- Pulmonary artery catheters are most useful in hypotensive patients. Their use is not without complication, however, and a definite indication for their use should be present. All patients with heart failure post-MI do not require a plumonary artery catheter.

Further reading

Fitzpatrick, A. and Sutton, R. (1992). A guide to temporary pacing. *Br. Med. J.*, **304**, 3665–9.
Ford, M. J., Robertson, C. E., and Munro, J. F. (1987). *Manual of medical procedures*. Churchill Livingstone, Edinburgh.
(1988). *Procedures in practice*. British Medical Journal, London.

NOTES

NOTES

NOTES

NOTES

CHAPTER 18

Useful information

Drug infusions

Isoprenaline

Indication

Beta-agonist that accelerates ventricular rate in a patient with symptomatic bradyarrhythmia while awaiting temporary pacing.

Drug regime

5 mg **isoprenaline** in 500 ml 5% dextrose to give a concentration of 10 μg/ml, infused at 1 ml per minute initially, increasing the infusion rate to maintain an adequate ventricular rate.

Dobutamine

Indication

Beta-agonist that increases cardiac output by stimulating the myocardium in a low output state.

Drug regime

500 mg **dobutamine** in 500 ml 5% dextrose to give a drug concentration of 1000 μg/ml. Can be given via a peripheral line. Infuse at 2.5 μg/kg/min initially, increasing to a maximum of 20 μg/kg/min. The following table gives infusion rates in ml/hour for a solution of 1000 μg/ml according to dose required and body weight.

Dobutamine infusion rate (μg/kg/min)	Body weight (kg)				
	50	60	70	80	90
2.5	7	9	10	12	13
5.0	15	18	21	24	27
10	30	36	24	48	54
15	45	54	63	72	81
20	60	72	84	96	108

Dopamine

Indication

At low (2.5 μg/kg/min) infusion rates **dopamine** acts selectively on renal dopamine receptors to dilate renal blood vessels and increase renal blood flow, and is therefore useful for oliguric patients with low cardiac output. At higher doses (above 5 μg/kg/min) dopamine acts on alpha receptors to produce vasoconstriction and may therefore be useful in a patient with persistent hypotension. **Dopamine** should only be given via a central line, since extravasation from a peripheral line may cause severe ischaemia due to vasoconstriction.

Drug regime

Add 800 mg **dopamine** to 500 ml 5% dextrose to give a drug concentration of 1600 μg/ml. Infuse at 2.5–5.0 μg/kg/min to produce optimal renal vasodilatation. Infuse at 5–20 μg/kg/min to support a low blood pressure by producing systemic vasoconstriction. The following table gives infusion rates in ml/hour for a 1600 μg/ml solution according to dose required and body weight.

Dobutamine infusion rate (μg/kg/min)	Body weight (kg)				
	50	60	70	80	90
2.5	5	6	7	8	9
5.0	10	12	14	16	18
10	20	23	27	31	35
15	29	35	41	47	53
20	39	47	55	62	70

Noradrenaline

Indication

Noradrenaline is a poworful alpha- and beta-agonist, and as such can be used to produce systemic vasoconstriction raising blood pressure in a severely hypotensive patient when other measures have failed. It is likely to produce tachycardia and provoke arrhythmias, and should only be used as a last resort.

Drug regime

5 mg **noradrenaline** in 500 ml 5% dextrose to give a concentration of 10 μg/ml, infused at 12 ml/hour initially, increasing to 120 ml/hour depending on response.

Sodium nitroprusside

Indication

Rapid control of elevated blood pressure by direct arterial

vasodilatation, in hypertensive emergencies or aortic dissection.

Drug regime

50 mg of **nitroprusside** in 500 ml of 5% dextrose to give a concentration of 100 μg/ml. Start infusion at 10 μg/min (6 ml/hour) and increase in steps of 10 μg/min every 5 minutes until desired blood pressure is achieved. **Nitroprusside** infusions need to be protected from direct exposure to light and should, ideally, only be used when an arterial line is in place to allow continuous direct measurement of arterial blood pressure.

Intravenous labetalol

Indication

Labetalol is an alpha- and beta-blocker, and is primarily used to control blood pressure in the treatment of hypertensive emergencies, or acute aortic dissection.

Drug regime

Make up a solution of 1 mg/ml by adding 20 ml of intravenous **labetalol**, containing 200 mg of drug, to 180 ml of 5% dextrose to make up a total volume of 200 ml. Commence infusion at 2 mg/min and increase every 15 minutes to a maximum of 200 mg/min, or until blood pressure falls to the desired level.

Intravenous nitrates

Indication

IV nitrates act by causing vasodilatation of coronary and systemic blood vessels. They act on the systemic vessels to produce venodilatation at low doses with added arterial dilatation at higher doses. Increased coronary flow and reduction of pre- and afterload are beneficial in both acute ischaemia and heart failure.

Nitrates can be used as an alternative to **nitroprusside** to produce emergency control of blood pressure.

Drug regimes

Intravenous GTN

- 50 mg of **GTN** in 50 ml of solution as supplied by the manufacturer.
- Commence infusion at 5 μg/min (0.3 ml/min), increasing by 5 μg/min every 15 minutes until the desired effect is obtained. The maximum dose of **GTN** is 200 μg/min.

Intravenous isosorbide dinitrate

- 50 mg of **ISDN** in 50 ml of solution as supplied by the manufacturer.
- Commence infusion at 1 mg/hour (1 ml/hour), increasing by 1 mg/hour every 15 minutes until the desired effect is obtained. The maximum dose of **ISDN** is 10 mg/hour.

Lignocaine

Indication

Standard Class I antiarrhythmic for suppression of ventricular arrhythmias.

Drug regime

Given as 100 mg bolus followed by infusion of 4 mg/min for 30 minutes, 2 mg/min for two hours, then 1 mg/min thereafter for 24 hours (500 mg **lignocaine** in 500 ml 5% dextrose giving a concentration of 1 mg/ml).

Disopyramide

Indication

Class I antiarrhythmic agent for use when lignocaine has failed.

Drug regime

Give 2 mg/kg bolus (to a maximum of 150 mg) followed by an infusion of 0.4 mg/kg for 24 hours (500 mg of **Disopyramide** in 500 ml 5% dextrose giving a concentration of 1 mg/ml).

Amiodarone

Indication

Class III antirrhythmic for the treatment of refractory ventricular or supraventricular arrhythmias.

Drug regime

Given as 5 mg/kg in 250 ml 5% dextrose over 10–60 minutes followed by an infusion of 1200 mg in 250 ml 5% dextrose over 24 hours.

Insulin infusion regime

50 U of short-acting insulin (i.e. **Actrapid** or **Humulin S**) in 50 ml normal saline via an infusion pump to give an insulin concentration of 1 U per ml.

BM stick	Insulin infusion rate (units/hr)
3.5–4.0	0
4.0–7.0	1
7.0–9.0	2
9.0–11.0	3
11.0–17.0	4
17.0–28.0	6

Oxygen therapy

High-concentration oxygen therapy

- Used for patients with no history of COPD, who are not at risk of CO_2 retention.

- 6 l/min via an MC mask will give 60% O_2, the maximum that can be achieved in a non-ventilated patient.

Controlled oxygen therapy

- Used in patients with COPD who are at risk of CO_2 retention during O_2 therapy. If high-concentration O_2 is administered to these patients, their hypoxic drive will be removed, ventilation will be suppressed, and dangerous CO_2 retention may occur.
- Give either 28% O_2 (4 l/min via a 28% **Ventimask**) **OR** 24% (2 l/min via a 24% Ventimask).
- If you suspect a patient may be a CO_2 retainer, check blood gases before starting O_2 therapy and tailor administered O_2 concentration accordingly. Check subsequent blood gases to ensure that CO_2 retention is not occurring.

Normal ranges for common blood tests

Values are approximate and cover both males and females—your own lab. ranges may differ slightly from those quoted below.

Arterial blood gases

pH	7.35–7.45 (35–45 μmol/l)
pO_2	10.0–13.5 kPa (75–100 mmHg)
pCO_2	4.5–6.0 kPa (35–45 mmHg)

Haematology

Hb	11–18 g/dl
WBC	4–11 × 10^a/l
PLATELETS	150–350 × 10^a/l
PCV	40–55%
MCV	76–100%
RETIC	0.2–2.0%
ESR	1–15 min/hr

Biochemistry

Na^+	135–145 mmol/l
K^+	3.5–5.0 mmol/l
Urea	2.5–6.5 mmol/l
Creatinine	55–150 mmol/l
Bicarbonate	24–30 mmol/l
Bilirubin	2–17 mmol/l
ALT	10–49 Units/l
AST	10–40 Units/l
Alk. phos.	40–100 Units/L
LDH	100–300 Units/l
Gamma GT	5–55 Units/l
Total protein	60–80 g/l
Albumin	35–50 g/l
Calcium	2.1–2.6 mmol/l
Fasting glucose	3.5–6.0 mmol/l
CPK	30–150 Units/l
Magnesium	0.075–1.0 mmol/l

NOTES

NOTES

NOTES

NOTES

General reading

Bennet, D. H. (1989). *Cardiac arrhythmias*. Wright, London.

Braunwald, E. (ed.) (1988). Heart disease. A text-book of cardiovascular medicine. W. B. Saunders, Philadelphia.

Hampton, J. R. (1986). The ECG made easy. Churchill Livingstone, Edinburgh.

Hamer, J. (ed.) (1987). Drugs for heart disease. Chapman and Hall, London.

Hampton, J. R. (1986). The ECG in practice. Churchill Livingstone, Edinburgh.

Jackson, G. (1990). Difficult cardiology. Blackwell Scientific Publications, London.

Julian, D. G., Camm, A. J., Fox, K. M., Hall, R. J. C., and Poole-Wilson, P. A. (ed.) (1989). *Diseases of the heart*. Baillère Tindall, London.

Opie, L. H. (1991). Drugs for the heart. W. B. Saunders, Philadelphia.

Schamroth, C. (1990). An introduction to electrocardiography. Blackwell Scientific Publications, London.

Swanton, R. H. (1990). Cardiology. Blackwell Scientific Publications, London.

Useful addresses

British Heart Foundation
14 Fitzhardinge Street
London
W1H 4DH
Tel. 071 935 0185

The Chest, Heart and Stroke Association
CHSA House
Whitecross Street
London
EC1 8IJ
Tel. 071 490 7999

APPENDIX: Algorithms for coronary care

A.1.1 Immediate treatment of acute MI

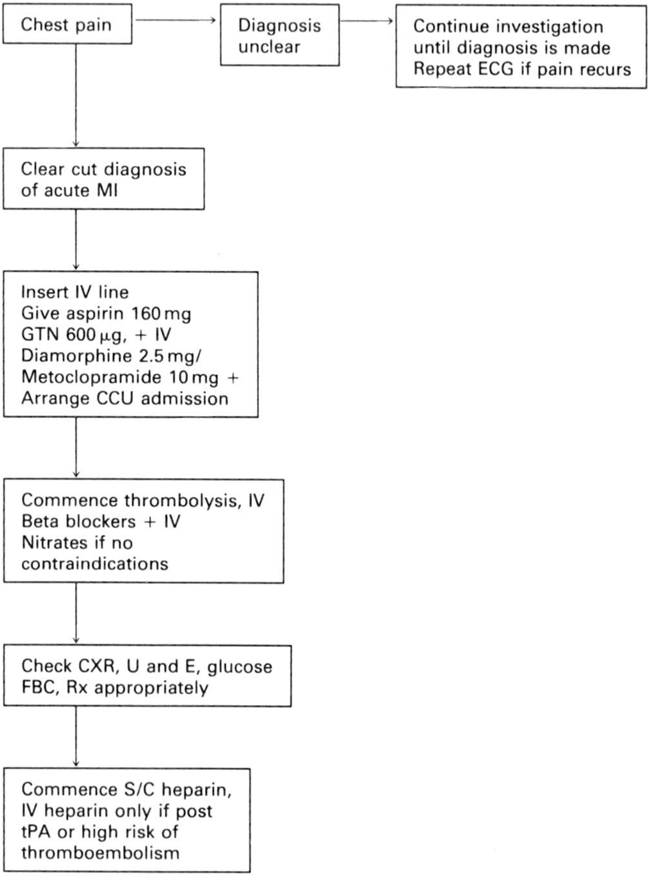

A.1.2 Management of uncomplicated acute MI after first 24 hours

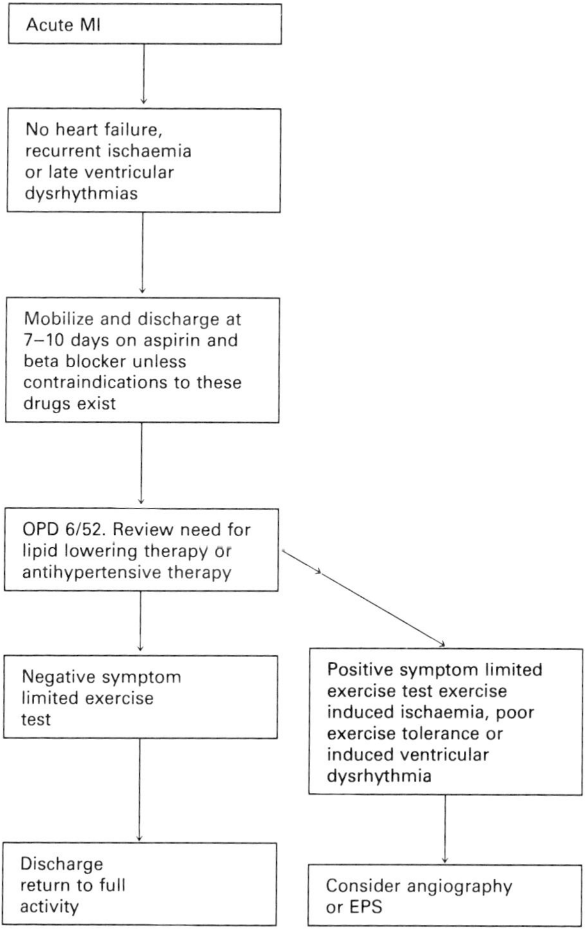

A.1.3 Administration of thrombolysis

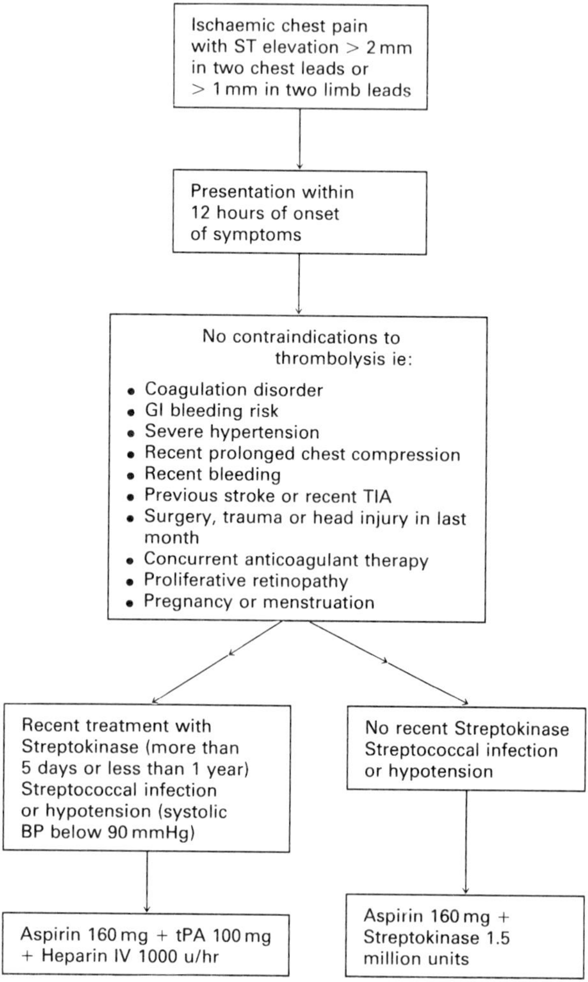

A.1.4 Administration of metoprolol

Ischaemic chest pain with ST elevation > 2 mm in two chest leads or > 1 mm in two limb leads within the last 12 hours

↓

Absence of:

- Bradycardia
- Hypotension
- heart block
- Concurrent Rx with beta blocker or verapamil
- Overt LVF
- Bronchospasm likely

↓

Give Metoprolol 5 mg IV

Wait 2 minutes

↓

Give Metoprolol 5 mg IV
If BP > 100, heart rate > 60 BPM
+PR interval < 0.26 s and no
aggravation of dyspnoea or sweating

↓

Wait 2 minutes

↓

Give Metoprolol 5 mg IV
If BP > 100, heart rate > 60 BPM
+PR interval < 0.26 s and no
aggravation of dyspnoea or sweating

↓

Wait 15 minutes

↓

Commence Metoprolol 50 mg qds for 48 hours if no contraindication has developed

A.1.5 Treatment of acute LVF

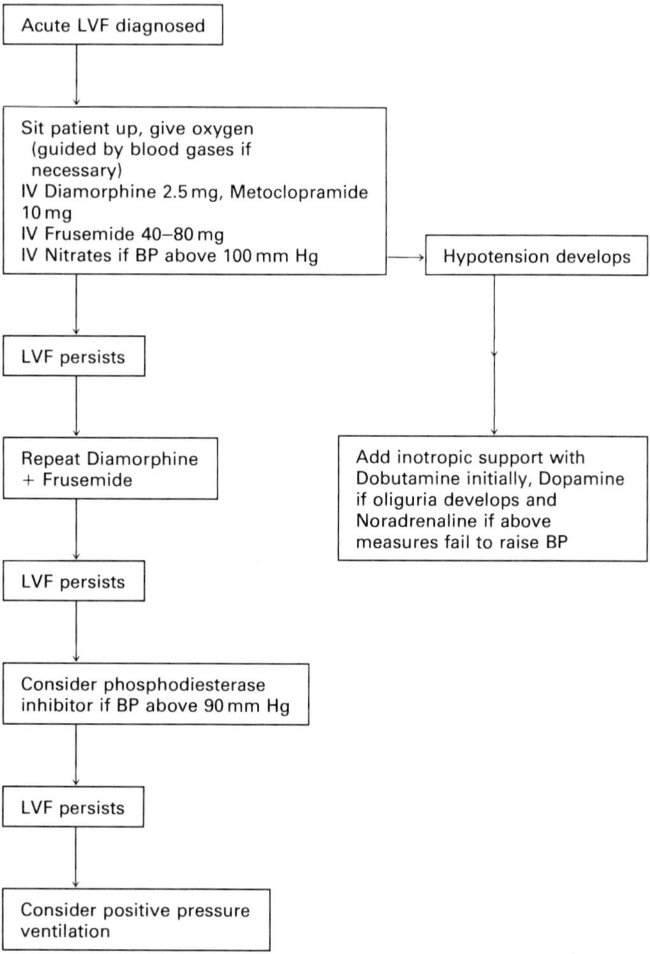

A.1.6 Treatment of cardiogenic shock

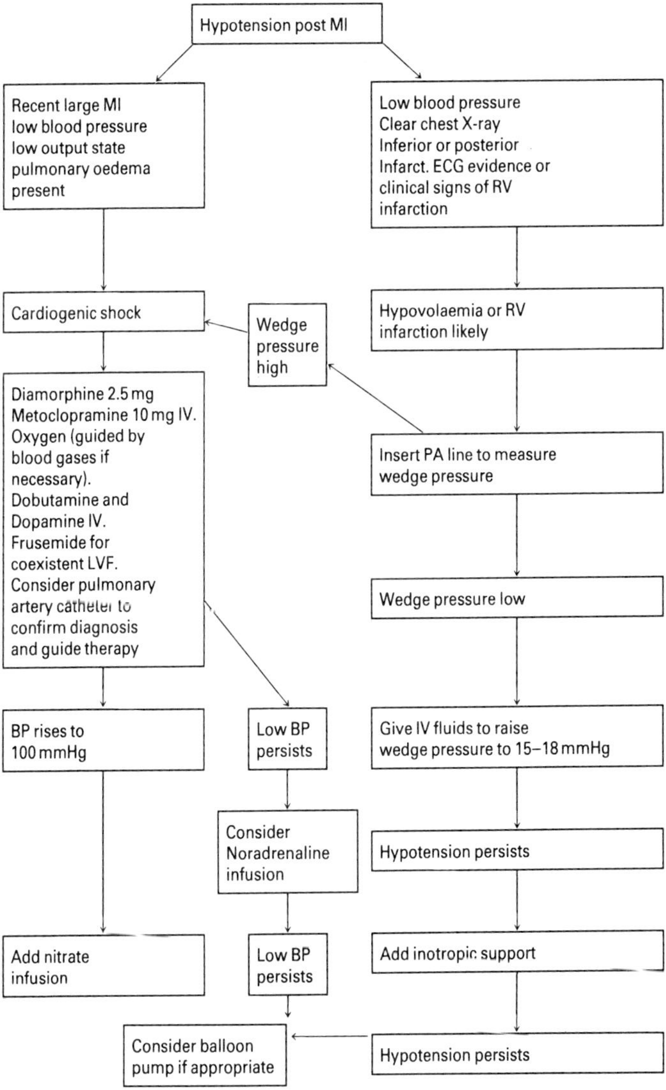

A.1.7 Ventricular tachycardia

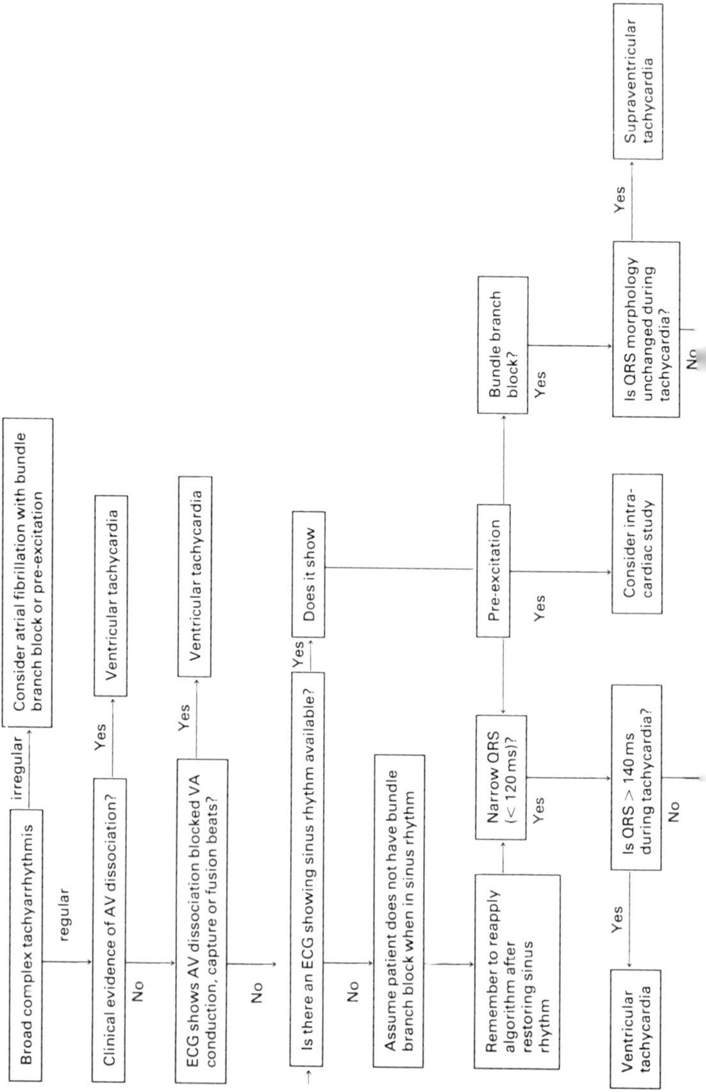

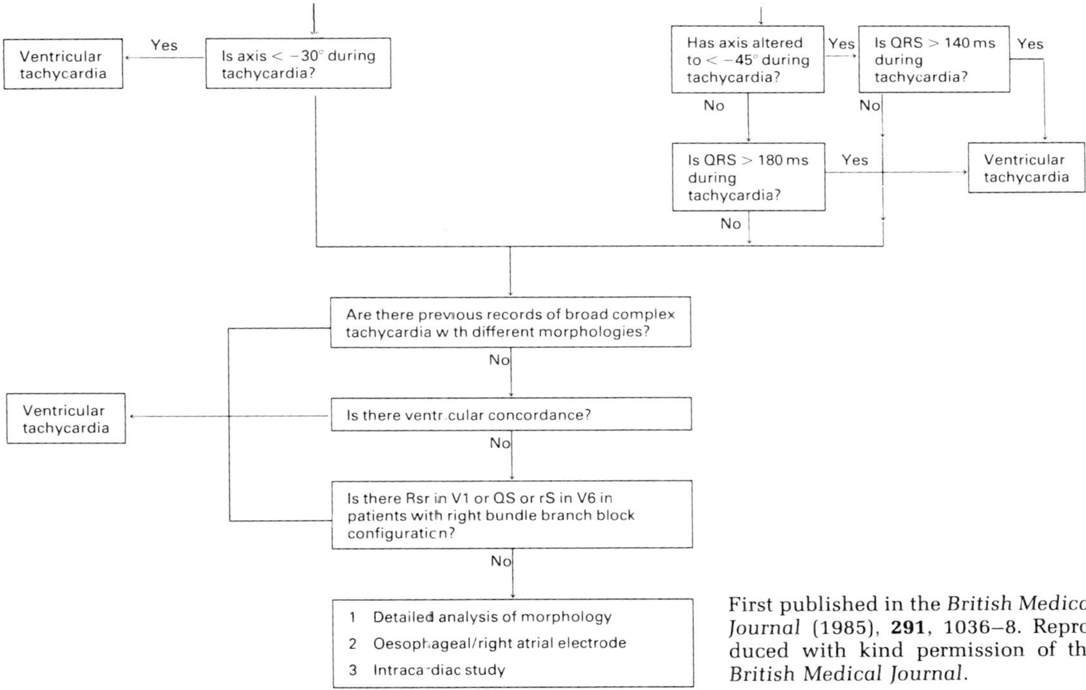

First published in the *British Medical Journal* (1985), **291**, 1036–8. Reproduced with kind permission of the *British Medical Journal*.

A.1.8 Treatment of VT

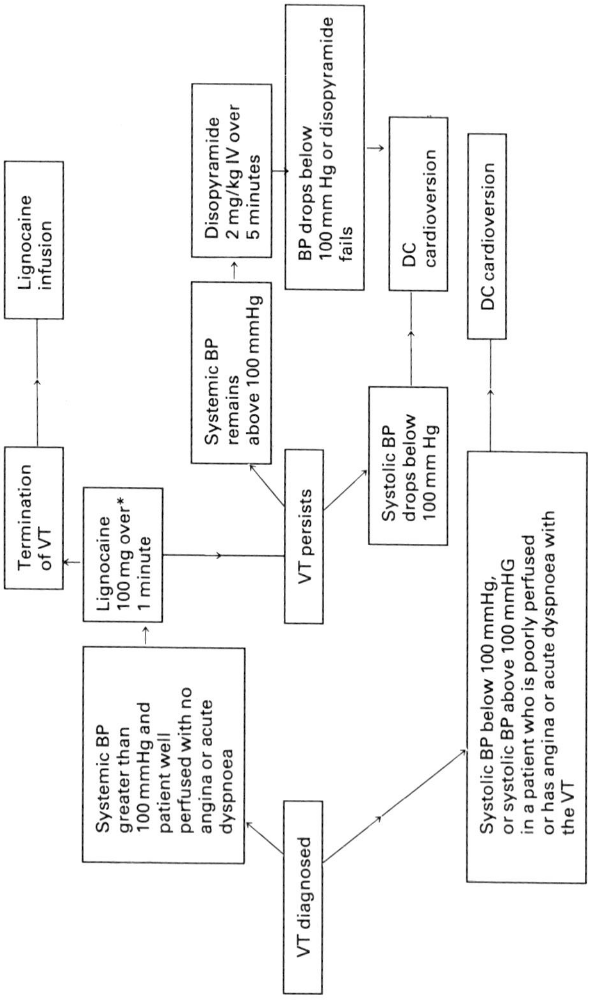

* Overdrive pacing can be attempted if a temporary pacing wire is already *in situ*—if this fails and the patient is stable try Lignocaine as above. If haemodynamic compromise develops, proceed to DC cardioversion.

A.1.9 Treatment of cardiac arrest due to VF: (remember to continue basic life support between shocks, and for 2 minutes after drug administration to allow circulation)

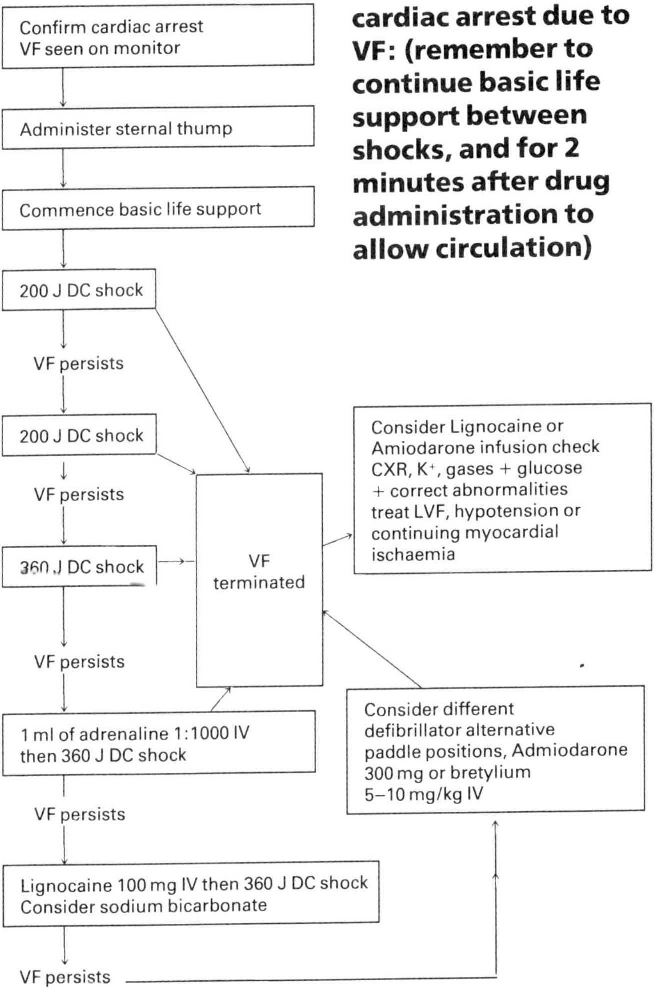

A.1.10 Treatment of Asystole/EMD (remember to continue basic life support for 2 minutes after drug administration to allow circulation)

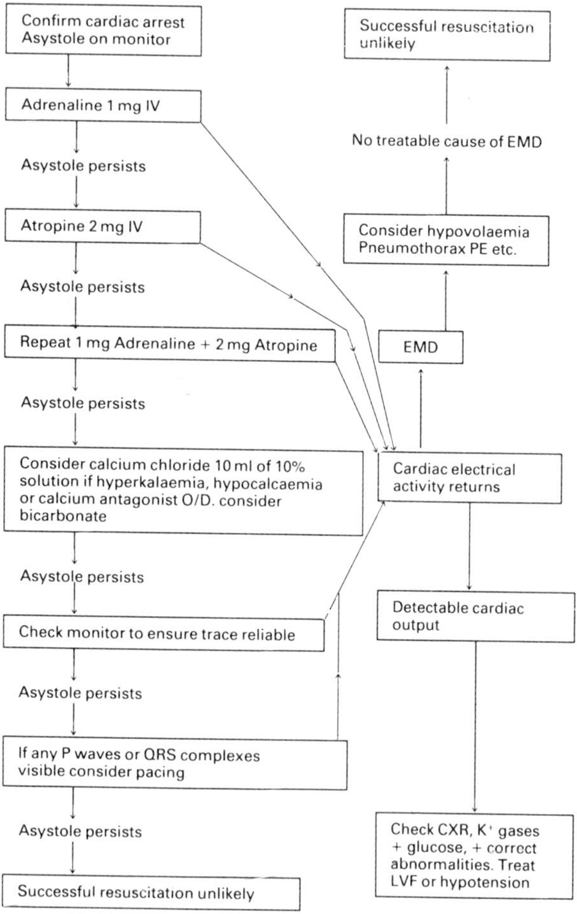

* Remember that VF may be present but not observed on the monitor —if you cannot exclude VF, administer 3 DC shocks (200 J, 200 J, and 360 J) before proceeding to adrenaline.

A.1.11 Temporary pacing post-MI

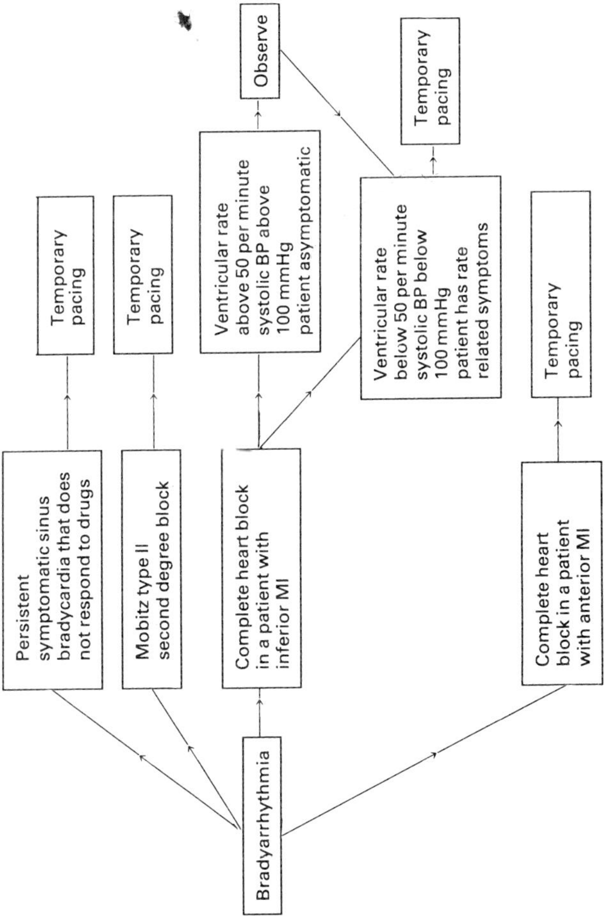

* Temporary pacing may also be indicated prophylactically in a patient who develops bundle branch block in the course of an acute myocardial infarction, particularly if this is associated with a change in electrical axis, or prolongation of the PR interval—policy varies among physicians, so seek senior advice in these patients.

A.1.12 Treatment of unstable angina

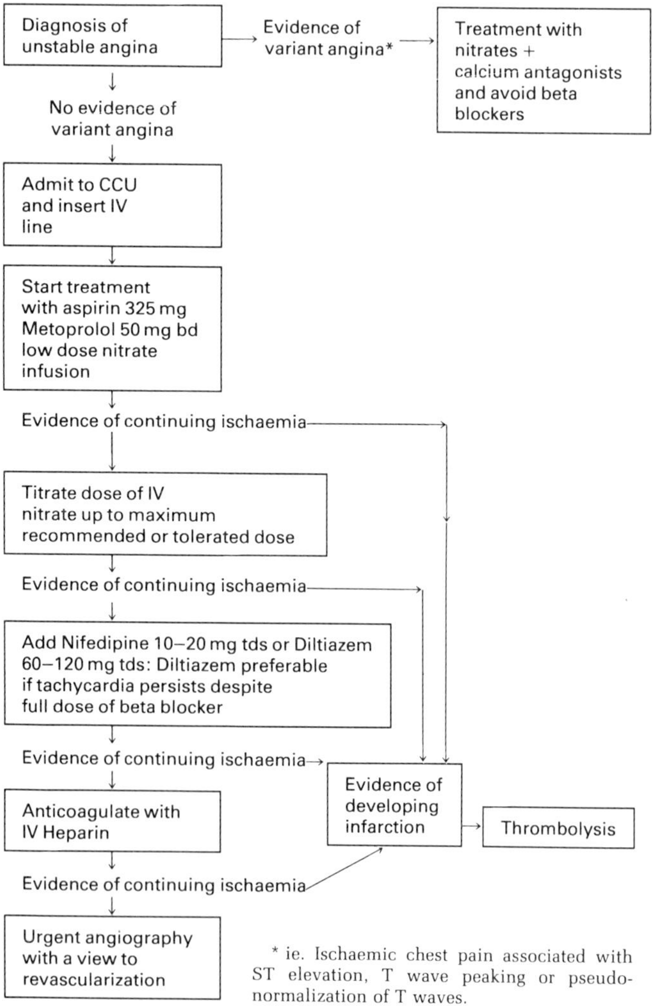

* ie. Ischaemic chest pain associated with ST elevation, T wave peaking or pseudo-normalization of T waves.

Index